AF342554

Herpetological Bibliography of Indonesia

Herpetological Bibliography of Indonesia

INDRANEIL DAS

KRIEGER PUBLISHING COMPANY
MALABAR, FLORIDA
1998

cover: mangrove cat snake (*Boiga dendrophila*).

Original Edition   1998

Printed and Published by
KRIEGER PUBLISHING COMPANY
KRIEGER DRIVE
MALABAR, FLORIDA 32950

FROM A DECLARATION OF PRINCIPLES JOINTLY ADOPTED BY A COMMITTEE OF THE AMERCAN BAR ASSOCIATION AND A COMMITTEE OF PUBLISHERS:
This publication is designed to provide accurate and authoritative information in regard to the subject matter covered. It is sold with the understanding that the publisher is not engaged in rendering legal, accounting, or other professional service. If legal advice or other expert assistance is required, the services of a competent professional person should be sought.

Library of Congress Cataloging-In-Publication Data

Das, Indraneil, 1964–
     Herpetological Bibliography of Indonesia / Indraneil Das.
          p.   cm.
     Includes bibliographical references.
     ISBN 1-57524-026-2 (hardcover : alk. paper)
     1. Reptiles—Indonesia—Bibliography. 2. Amphibians—Indonesia—
Bibliography. 3. Herpetology—Indonesia—Bibliography. I. Title.
Z7998.I5D37   1998
[QL661.I5]
016.5979′09598—dc21   97-50219
                                                                              CIP

10 9 8 7 6 5 4 3 2

INTRODUCTION

The archipelago of Indo-Malaya, stretching from the Andaman and Nicobar Islands, eastward to Timor, is the largest in the world (approximately 14,000 islands), and includes several political entities. Much of the land area and associated coastal water fall within the political boundaries of a single nation, that of the Republic of Indonesia. However, significant portions are controlled by India (most of the Andaman Islands and all of the Nicobars), Myanmar or Burma (the Cocos Islands of the Andamans), Malaysia (Sarawak and Sabah) and the Philippines (Palawan and associated islands that sit on the Sunda shelf). Although these appear to be a highly heterogeneous assemblage of islands randomly selected, this is also the coverage of *The Reptiles of the Indo-Australian Archipelago* by Nellie De Rooij (1915; 1917), except the present work excludes New Guinea and associated islands.

This comprehensive bibliography includes all scientific papers, magazine and newspaper articles, books, book reviews, field/excursion reports, museum catalogues, bibliographies, chapters from books and theses known to me that contain references to any species of amphibians and reptiles that occur in the region. Both terrestrial and aquatic taxa (including recent and fossil forms) are included. Languages in which the literature on the herpetology of the region has appeared include Bahasa Indonesia, Chinese, Dutch, English, French, German, Hungarian, Italian, Japanese, Latin, Polish, Portuguese, Russian and Spanish. The bias towards literature on faunistics and natural history, rather than captive husbandry and studies on snake bite and venom, will be obvious.

Library work for this bibliography was conducted at the Sarawak Museum, Kuching, Malaysia; Brunei Museum, Bandar Seri Begawan, Brunei Darussalam; University Brunei Darussalam, Bandar Seri Begawan, Brunei Darussalam; University of the Philippines, Los Banõs, Philippines; Natural History Museum, London, United Kingdom; Museum National d'Histoire Naturelle, Paris, France; Centre for Herpetology, Madras Crocodile Bank Trust, Madras, India; Field Museum of Natural History, Chicago, USA; National Museum of Natural History, Smithsonian Institution, Washington, DC, USA, Ernst Mayr (Museum of Comparative Zoology) and Widener Libraries of Harvard University, Cambridge,

Mass., USA, as well as the personal library of Van Wallach, Cambridge, Mass., USA. The cut-off date for inclusion of entries was October 22, 1997. No doubt there will be omissions, and additional references will be included in future editions of this work.

I thank (in alphabetic sequence): Harry V. Andrews, E. Nicholas Arnold, John E. Cadle, Ronald I. Crombie, Alain Dubois, W. Ronald Heyer, Robert F. Inger, Charles Leh, Colin J. McCarthy, Alan Resetar, José Rosado, Harold K. Voris, Van Wallach, Romulus Earl Whitaker, Ernst E. Williams and George R. Zug, for facilities at their respective institutions.

BIBLIOGRAPHY

ABDULALI, H. 1971. Narcondam Island and notes on some birds from the Andaman Islands. *J. Bombay nat. Hist. Soc.* 68(2): 385–411.

ABDULALI, H. 1982. Some field notes on the newly described toad, *Bufo camortensis* Mansukhani & Sarkar. *J. Bombay nat. Hist. Soc.* 79(2): 430.

ADLER, K. 1989. Herpetologists of the past. *In*: Contributions to the history of herpetology. pp: 5–141. K. Adler (Ed.). Contributions to Herpetology, Number 5, Oxford, Ohio.

ADLER, K. & E.-M. ZHAO. 1994. The proper name for the Asian wolf snake: *Lycodon* (Serpentes: Colubridae). *Sichuan J. Zool.* 14(2): 74–75.

AHL, E. 1926. Neue Eidechsen und Amphibien. *Zool. Anz.* 47(7/8): 186–192.

AHL, E. 1927. Zur Systematik der Asiatischen Arten der Fröschgattung *Rhacophorus. Sitz. Ges. Natur. Freunde, Berlin* 1927: 35–47.

AHL, E. 1927. Ueber neue oder seltene Fröschlurche aus dem Zoologische Museum Berlin. *Sitz. Ges. Natur. Freunde, Berlin* 1927: 111–117.

AHL, E. 1929. Beschreibung eines neuen Laubfrösches der Gattung *Hyla* von Java. *Zool. Anz.* 85(9/10): 269–271.

AHL, E. 1931. Anura III. Das Tierreich 55. Walter de Gruyter & Co. XVI + 477 pp.

AHL, E. 1933. Ergebnisse der Celebes und Halmahera Expedition Heinrich 1930–32. 1. Reptilien und Amphibien. *Mitt. Zool. Mus. Berlin* 19: 577–583.

AIKANATHAN, S. 1996. Malaysian sea turtles: what future? *Mar. Turtle Newsl.* (73): 23–24.

ALCALA, A. C. 1986. Guide to Philippine flora and fauna. Volume X. Amphibians and reptiles. Natural Resources Management Centre, Ministry of Natural Resources/University of the Philippines, Manila. xiv + 195 pp.

ALCALA, A. C., G. JOERMANN & J. BRZOSKA. 1986. Calls of certain Philippine anurans (Microhylidae, Ranidae). *Silliman J.* 33(1–4): 31–44.

ANANJEVA, N. B. & T. N. MATVEYEVA-DUJSEBAYEVA. 1996. Some evidence of *Gonocephalus* species complex divergence basing on skin sense organ morphology. *Russian J. Herpetol.* 3(1): 82–88.

ANDERSON, J. A. R., A. C. JERMY & THE EARL OF CRANBROOK. 1982. Gunung Mulu National Park. Royal Geographical Society, London. 345 pp.

ANDERSSON, L. G. 1899. Catalogue of Linnean type-specimens of snakes in the Royal Museum in Stockholm. *Bihang Till K. Sevnska Vet.-Akad. Handlingar* 24(6): 3–35.

ANDERSSON, L. G. 1900. Catalogue of Linnean type-specimens of Linnaeus's Reptilia in the Royal Museum in Stockholm. *Bihang Till K. Sevnska Vet.-Akad. Handlingar* 26(1): 3–29.

ANDERSSON, L. G. 1923. Some reptiles and batrachians from central Borneo. *Med. Fra. det. Zool. Mus. Kristiana* 7: 120–125.

ANDREWS, H. V. 1997. Population dynamics and ecology of the saltwater crocodile (*Crocodylus porosus* Schneider) in the Andaman and Nicobar Islands. Interim report. Phase III. Submitted to the Andaman and Nicobar Islands Forest Department and the Centre for Herpetology, Madras Crocodile Bank Trust (AN/C-3/97). 6 pg.

ANDREWS, H. V. & R. WHITAKER. 1994. Population dynamics and ecology of the saltwater crocodile (*Crocodylus porosus*) (Schneider) in the Andaman and Nicobar Islands. Interim survey report. Phase I. Submitted to the Andaman and Nicobar Forest Department, Asian Wetland Bureau and the Centre for Herpetology (AN/C-1-94). 33 pg.

ANDREWS, H. V. & R. WHITAKER. 1994. Population dynamics and ecology of the saltwater crocodile (*Crocodylus porosus*) (Schneider) in the Andaman and Nicobar Islands. Interim survey report. Phase II. Submitted to the Andaman and Nicobar Forest Department and the Centre for Herpetology (AN/C-2-94). 18 pg.

ANDREWS, H. V. & R. WHITAKER. 1994. Status of the saltwater crocodile (*Crocodylus porosus* Schneider, 1801) in North Andaman Island. *Hamadryad* 19: 79–92.

ANNANDALE, J. & C. FLETCHER. 1988. *Gonyosoma oxycephala. Snake Keeper* 2(5): 9–10.

ANNANDALE, N. 1904. Contributions to Oriental herpetology I—The lizards of the Andamans, with the description of a new gecko and a note on the reproduced tail in *Ptychozoon homalocephalum. J. Asiatic Soc. Bengal* 63: 12–22.

ANNANDALE, N. 1905. Contributions to Oriental herpetology II—Notes on the Oriental lizards in the Indian Museum, with a list of the species recorded from British India and Ceylon. Part I. *J. & Proc. Asiatic Soc. Bengal* n.s. 1(3): 81–93; Pl. I–II.

ANNANDALE, N. 1905. Contributions to Oriental herpetology III—Notes on the Oriental lizards in the Indian Museum, with a list of the species recorded from British India and Ceylon. Part 2. *J. & Proc. Asiatic Soc. Bengal* n.s. 1(3): 139–151.

ANNANDALE, N. 1905. Additions to the collection of Oriental snakes in the Indian Museum. -Part II.—Specimens from the Andamans and Nicobars. *J. Asiatic Soc. Bengal* 64: 173–176.

ANNANDALE, N. 1913. Some new and interesting batrachian and lizards from India, Ceylon and Borneo. *Rec. Indian Mus.* 9(20): 301–307; Pl. XV.

ANNANDALE, N. 1917. Report on a collection of reptiles and batrachians from Java. *J. Fed. Malay Mus.* 7: 107–111.

ANNANDALE, N. 1917. Zoological results of a tour in the Far East. Batrachia. *Mem. Asiatic Soc. Bengal* 6: 119–155; Pl. V–VI.

ANONYMOUS. 1927. Over het verslinden van menschen door *Python reticulatus. Trop. Natuur* 16: 65.

ANONYMOUS. 1941. De maaltijd van een python. *Ned.-Ind. Jager* 11: 46.

ANONYMOUS. 1978. Sabah protects its living dinosaurs. *Malayan Natural.* 32(2): 8–9.

ANONYMOUS. 1984. Sea turtle trade in Indonesia. Report to the Directorate of Nature Conservation, Directorate-General of Forest Protection and Nature Conservation (PHPA), Bogor. 56 pg.

ANONYMOUS. 1992. Turtle business obscures protection efforts in Bali. *Mar. Turtle Newsl.* (57): 28–29.

APPLEMAN, F. J. 1934. *Crocodylus porosus. Trop. Natuur* 23: 228.

ARNDT, J. W. E. 1859. Reptiliën van Wonsobo in Midden-Java. *Natuur. Tijd. Ned-Indië* 16(2): 301–302.

ARNDT, J. W. E. 1859. Reptiliën van Midden-Java en Manado. *Natuur. Tijd. Ned-Indië* 16(2): 314.

ASHE, J. S. & H. MARX. 1988. Phylogeny of the viperine snakes (Viperinae): Part II. Cladistic analysis and major lineages. *Fieldiana Zool. n.s.* (52): i–iii + 1–23.

ASHTON, P. 1963. (a) Dr. Ashton's own story. *In:* Snake bites man: two recent Borneo cases. Complied by N. S. Haile. *Sarawak Mus. J.* 11: 291–298.

ATTENBOROUGH, D. 1957. Zoo quest for a dragon. Lutterworth Press, London. 157 pp.

AUDY, J. R. 1953. Strolling on the ceiling. *Malay nat. J.* 7: 182–190.

AUFFENBERG, W. 1970. A day with No. 19—Report on a study of the Komodo monitor. *Anim. Kingdom* 6: 18–23.

AUFFENBERG, W. 1971. We lived with dragons. *Sci. Digest* August 1971: 32–38.

AUFFENBERG, W. 1972. Komodo dragons. *Nat. Hist.* 81(4): 52–59.

AUFFENBERG, W. 1974. Checklist of fossil land tortoises (Testudinidae). *Bull. Florida St. Mus.* 18(3): 121–251.

AUFFENBERG, W. 1978. Social and feeding behavior in *Varanus komodoensis. In*: Behavior and neurobiology of lizards. pp: 301–331. N. Greenberg & P. D. McLean (Eds.). Government Printing Office, Washington, D.C.

AUFFENBERG, W. 1979. Detective Auffenberg gets his beast. *Int. Wildl.* 9(2): 12–15.

AUFFENBERG, W. 1980. The herpetofauna of Komodo, with notes on adjacent areas. *Bull. Florida State Mus., Biol. Sci.* 25: 39–156.

AUFFENBERG, W. 1981. The behavioral ecology of the Komodo monitor. University Presses of Florida, Gainesville. x + 406 pp.

AUTH, D. L., K. AUFFENBERG & D. K. DORMAN. 1990. Geographical distribution. *Gerarda prevostiana* (Gerard's water snake). *Herpetol. Rev.* 21(2): 41.

BACON, J. P. 1967. Systematic status of three scincid lizards (genus *Sphenomorphus*) from Borneo. *Fieldiana Zool.* 51: 63–76.

BANKS, E. 1937. The breeding of the edible turtle (*Chelonia mydas*). *Sarawak Mus. J.* 4: 523–532.

BARBE, P. 1846. Notes of Nicobar Islands. *J. Asiatic Soc. Bengal* 15: 344–366.

BARBOUR, T. 1903. A new species of flying lizard from Sarawak, Borneo. *Proc. Biol. Soc. Washington.* 16: 59–60.

BARBOUR, T. 1904. A new batrachian from Sarawak, Borneo. *Proc. Biol. Soc. Washington* 17: 51–52.

BARBOUR, T. 1911. New lizards and a new toad from the Dutch East Indies, with notes on other species. *Proc. Biol. Soc. Washington* 24: 15–22.

BARBOUR, T. 1912. A contribution to the zoogeography of East Indian Islands. *Bull. Mus. Comp. Zool., Harvard Coll.* 44: 1–203 + 8 pl.

BARBOUR, T. 1923. Reptiles in the East and West Indies and some digression. *American Nat.* 57: 125–128.

BARBOUR, T. 1927. Two new Bornean snakes. *Proc. Biol. Soc. Wash.* 40: 127–128.

BARBOUR, T. 1938. Notes on "*Nectophryne*". *Proc. Biol. Soc. Washington* 51: 191–195.

BARBOUR, T. & G. K. NOBLE. 1916. New amphibians and a new reptile from Sarawak. *Proc. New England Zool. Club* 6: 19–22.

BARCLAY, J. 1988. A stroll through Borneo. January Books Ltd., Wellington. Second edition.

BARKER, D. & T. BARKER. 1994. The blood python and other subspecies of the short-tailed python (*Python curtus*). *The Vivarium* 6(3): 30–33; 35.

BARR, C. 1991. Current status of trade and legal protection for sea turtles in Indonesia. *Mar. Turtle Newsl.* (54): 4–7.

BARTLETT, E. 1895. The crocodiles and lizards of Borneo in the Sarawak Museum, with descriptions of supposed new species, and the variation of colours in the several species during life. *J. Str. Br. Royal Asiatic Soc.* 1895(28): 73–96.

BARTLETT, E. 1895. Notes on the batrachians, or frogs and toads of Borneo and the adjacent islands. *Sarawak Note Book* (1): 7–14.

BARTLETT, E. 1895–1896. Notes on the snakes of Borneo and the adjacent islands. Parts I–III. *Sarawak Note Book* (1): 68–85; (2): 100–116.

BAUER, A. M. 1994. On the identity of *Euprepes samoensis moluccensis* Peters, 1864. *J. Herpetol.* 28(2): 257–258.

BAUER, A. M. & R. GÜNTHER. 1991. An annotated type catalogue of the geckos (Reptilia: Gekkonidae) in the Zoological Museum, Berlin. *Mitt. Zool. Mus. Berlin* 67(2): 279–310.

BAUER, A. M. & R. GÜNTHER. 1995. An annotated type catalogue of the lacertids (Reptilia: Lacertidae) in the Zoological Museum, Berlin. *Mitt. Zool. Mus. Berlin* 71(1): 37–62.

BAUER, A. M., R. GÜNTHER & M. KLIPFEL. 1995. The herpetological contributions of Wilhelm C. H. Peters (1815–1883). Society for the Study of Amphibians and Reptiles/Deutsche Gesselschaft für Herpetologie und Terrarienkunde, Facsimile Reprints in Herpetology, Oxford, Ohio. 714 pp.

BAUER, A. M. & K. HENLE. 1994. Familia Gekkonidae (Reptilia, Sauria) I. Australia and Oceania. Das Tierreich 109. Walter de Gruyter, Berlin and New York. xiii + 306 pp.

BAUER, J. R. A. 1860. Reptiliën van Makassar. *Natuur. Tijd. Ned-Indië* 22(2): 80–81.

BAUMANN, F. 1913. Reptilien und Batrachier des Berner Naturhistoirischen Museums aud dem Battak-Gebirge von West Sumatra. *Zool. Jb. Syst.* 34: 257–278.

BAYLEY-DE-CASTRO, A. 1927. A case of snake-bite due to Cantor's viper (*Lachesis cantoris*). *J. Bombay nat. Hist. Soc.* 32(1): 223–224.

BECCARI, O. 1904. Wanderings in the great forests of Borneo. Archibald Constable & Co., Ltd., London. Reprinted 1989, Oxford University Press, Singapore. 424 pp.

BEDOT, M. 1909. Sur la faune de l'Archipel Malais. *Rev. Suisse Zool.* 17(1): 143–169.

BELL, T. 1834. Characters of a new genus of freshwater tortoise (*Cyclemys*). *Proc. Zool. Soc. London* 1834: 17.

BELT, P., A. MALHOTRA, R. S. THORPE, D. A. WARRELL & W. WIÜSTER. "1997" 1996. Russell's viper in Indonesia: snakebite and systematics. *In*: Venomous snakes: ecology, evolution and snakebite. pp: 219–234. R. S. Thorpe, W. Wüster & A. Malhotra (Eds.). Symp. Zool. Soc. London (70). Clarendon Press, Oxford.

BENNETT, D. 1996. Warane der Welt. Welt der Warane. Edition Chimaira, Frankfurt am Main. 383 pp.

BENNETT, J. 1988. Red-tailed tree racer (*Gonyosoma oxycephala*). *Snake Keeper* 2(1): 12–13.

BENTLEY, M. P. 1975. A fishing chichak? *Malayan Natural.* 1(3): 3.

BERGMAN, A. H. 1943. The breeding habits of sea snakes. *Copeia* 1943(3): 156–160.

BERGMAN, R. A. M. 1938. Een nieuw orgaantje bij zeeslangen. *Geneesk. Tijdschr. Ned-Indië* 78: 2061–2070.

BERGMAN, R. A. M. 1941. Het vervellen van zeeslangen en van *Acrochordus. Trop. Natuur.* 30: 145–148.

BERGMAN, R. A. M. 1941. *Acrochordus javanicus* Hornst. *Treubia* 18(2): 207–211.

BERGMAN, R. A. M. 1942. *Enhydrina schistosa. Natur. Tijdschr. Ned-Indië* 102(1): 9–12.

BERGMAN, R. A. M. 1943. The breeding habits of sea snakes. *Copeia* 1943: 156–159.

BERGMAN, R. A. M. 1951. The anatomy of *Homalopsis buccata. Proc. Koninkl. Ned. Akad. Wet. Amsterdam C* 54: 511–524.

BERGMAN, R. A. M. 1952. L'anatomie du genre *Ptyas* à Java. *Riv. Biol. Colon, Rome* 12: 1–42.

BERGMAN, R. A. M. 1953. The anatomy of *Cylindrophis rufus* (Laur.). *Proc. Koninkl. Ned. Akad. Wet. Amsterdam C* 56: 650–666.

BERGMAN, R. A. M. 1955. L'anatomie de *Cerberus rhynchops. Arch. Néerl. Zool.* 11: 113–126.

BERGMAN, R. A. M. 1955. L'anatomie de *Enhydrina schistosa. Arch. Néerl. Zool.* 11: 127–142.

BERGMAN, R. A. M. 1955. *Dendrelaphis pictus. Proc. Koninkl. Ned. Akad. Wet. Amsterdam C* 58: 206–218.

BERGMAN, R. A. M. 1956. The anatomy of *Dryophis prasinus. Proc. Koninkl. Ned. Akad. Wet. Amsterdam C* 59: 263–279.

BERGMAN, R. A. M. 1958. The anatomy of the Acrochordinae. *Proc. Koninkl. Ned. Akad. Wet. Amsterdam C* 61: 145–184.

BERGMAN, R. A. M. 1961. The anatomy of some Viperidae. *Acta Morphol. Neerl.-Scand.* 4: 195–230.

BERNELOT MOENS, J. G. T. 1859. Reptiliën en visschen van Batjan. *Natuur. Tijd. Ned-Indië* 16(2): 208.

BERNELOT MOENS, J. G. T. 1859. Reptiliën en visschen van Anjer. *Natuur. Tijd. Ned-Indië* 16(2): 423.

BERRY, P. Y. 1975. The amphibian fauna of peninsular Malaysia. Tropical Press, Kuala Lumpur. 130 pp.

BERRY, P. Y. & J. R. HENDRICKSON. 1963. *Leptobrachium nigrops*, a new pelobatid frog from the Malay Peninsula, with remarks on the genus *Leptobrachium* in southeast Asia. *Copeia* 1963: 643–648.

BEUTELSCHIESS, J. & C. B. BEUTELSCHIESS 1990. Zur kenntnis von *Leptobrachium hendricksoni* Taylor 1962. *Sauria, Berlin* 12(1): 3–10.

BEZUIJEN, M. R., P. CANNUCCIARI, C. MANOLIS, (?) RHIZA, (?) SAMEDI & B. K. SIMPSON. 1995. Field expedition to the Lalan River and its tributaries, south Sumatra, Indonesia, August-October 1995: Assessment of the distribution, abundance, status and nesting biology of the false gharial (*Tomistoma schlegelii*). Report to the IUCN/SSC Crocodile Specialist Group. 101 pg.

BHASKAR, S. 1979. Sea turtles in the South Andaman Islands. *Hamadryad* 4(1): 3–6.

BHASKAR, S. 1979. Sea turtle survey in the Andaman and Nicobars. *Hamadryad* 4(3): 2–26.

BHASKAR, S. 1980. Sea turtle surveys of Great Nicobar and Little Andaman Islands. Report to the WWF-I. 5 pg.

BHASKAR, S. 1981. Travels in the Andaman and Nicobar Islands—1979. *Hamadryad* 6(1): 2–8.

BHASKAR, S. 1984. Sea turtles in North Andaman and other Andaman islands. Report to the WWF-I. 46 pg.

BHASKAR, S. 1984. Locating and conserving sea turtle nesting grounds in the Andamans. *In*: Spirit of Enterprise: The 1984 Rolex Awards. pp: 190–193. Aurum Press, London.

BHASKAR, S. 1984. The distribution and status of sea turtles in India. *In*: Proc. Workshop Sea Turtle Conserv. pp: 21–35. E. G. Silas (Ed.). Central Marine Fisheries Research Institute Spec. Publ. No. 18. CMFRI, Cochin.

BHASKAR, S. 1993. The status and ecology of sea turtles in the Andaman and Nicobar Islands. Centre for Herpetology, Madras Crocodile Bank Trust. Publication No. ST 1/93. 37 pg.

BHASKAR, S. 1996. Renesting intervals of the hawksbill turtle (*Eretmochelys imbricata*) on South Reef Island, Andaman Islands, India. *Hamadryad* 21: 19–22.

BHASKAR, S. 1996. Sea kraits on South Reef Island, Andaman Islands, India. *Hamadryad* 21: 27–35.

BHASKAR, S. & H. A. ANDREWS. 1993. Action Plan for sea turtles in the Andaman and Nicobar Islands, India. *Mar. Turtle Newsl.* (60): 23.

BHASKAR, S. & G. C. RAO. 1992. Present status of some endangered animals in Nicobar Islands. *J. Andaman Sci. Assoc.* 8: 181–186.

BHASKAR, S. & R. WHITAKER. 1983. Sea turtle resources in the Andamans. *In*: Mariculture potential in Andaman and Nicobar Islands—An indicative survey. *Bull. Cent. Mar. Fish. Res. Inst.* 34: 94–97.

BISWAS, S. 1985. Some notes on the reptiles of the Andaman and Nicobar Islands. *J. Bombay nat. Hist. Soc.* 81: 476–481.

BISWAS, S. & D. P. SANYAL. 1965. A new species of wolf-snake of the genus *Lycodon* Boie [Reptilia: Serpentes: Colubridae] from the Andaman and Nicobar Islands. *Proc. Zool. Soc. Calcutta* 18: 137–141.

BISWAS, S. & D. P. SANYAL. 1977. Notes on the Reptilia collection from the Great Nicobar Island during the Great Nicobar Expedition in 1966. *Rec. Zool. Surv. India* 72: 107–124.

BISWAS, S. & D. P. SANYAL. 1977. A new species of skink of the genus *Dasia* Gray, 1889 (Reptilia: Scincidae). *J. Bombay nat. Hist. Soc.* 74: 133–136.

BISWAS, S. & D. P. SANYAL. 1978. A new species of krait of the genus *Bungarus* Daudin, 1803 [Serpentes: Elapidae] from the Andaman Island. *J. Bombay nat. Hist. Soc.* 75: 179–183

BISWAS, S. & D. P. SANYAL. 1980. A report on the Reptilia fauna of Andaman and Nicobar Islands in the collection of the Zoological Survey of India. *Rec. Zool. Surv. India* 77: 255–292.

BLANFORD, W. T. 1881. On a collection of reptiles and frogs chiefly from Singapore. *Proc. Zool. Soc. London* 1881: 215–226.

BLEEKER, P. 1850. Iets over kop van *Crocodylus* (*Gavialis*) *schlegelii* S. Müll. *Natuur. Tijd. Ned-Indië* 1: 313–314.

BLEEKER, P. 1857. Over eenige reptiliën van Celebes. *Natuur. Tijd. Ned-Indië* 14: 231–233.

BLEEKER, P. 1857. Over eenige reptiliën van het eiland Banka. *Natuur. Tijd. Ned-Indië* 14(4): 233–235.

BLEEKER, P. 1857. Berigt omtrent eenige reptiliën von Sumatra, Borneo, Batjan en Boero. *Natuur. Tijd. Ned-Indië* 13(3): 470–475.

BLEEKER, P. 1857. Opsomming der soorten van reptiliën, tot dus verre van het eiland Java bekend geworden. *Natuur. Tijd. Ned-Indië* 14(4): 235–244.

BLEEKER, P. 1858. Opsomming der tot dus verre het eiland Sumatra bekend gewordene reptiliën. *Natuur. Tijd. Ned-Indië* 15(1): 260–263.

BLEEKER, P. 1859. Eene verzameling reptilien en visschen van Sinkawang (wetkust van Borneo). *Natuur. Tijd. Ned-Indië* 16: 188–189.

BLEEKER, P. 1859. *Hemidactylus ludekingii* Blkr. *Natuur. Tijd. Ned-Indië* 16(2): 27.

BLEEKER, P. 1859. Visschen van Tikoe en reptiliën van Agam, aangeboden door E. W. A. Ludeking. *Tijd. Ned-Indië* 16(2): 26–28.

BLEEKER, P. 1859. *Hemidactylus platurus* Blkr. *Natuur. Tijd. Ned-Indië* 16(2): 31.

BLEEKER, P. 1859. Slangen van Fort de Kock, aangeboden dooe E. W. A. Ludeking. *Tijd. Ned-Indië* 16(2): 241.

BLEEKER, P. 1859. *Gonyosoma jansenii* Blkr., eene nieuwe slang van Manado. *Natuur. Tijd. Ned-Indië* 16(2): 242.

BLEEKER, P. 1859. *Bufo gymnauchen* Blkr. *Natuur. Tijd. Ned-Indië* 16(2): 46–47.

BLEEKER, P. 1859. *Hemidactylus meijeri* Blkr. *Natuur. Tijd. Ned-Indië* 16(2): 47.

BLEEKER, P. 1859. *Aipysurus margaritophorus*. *Natuur. Tijd. Ned-Indië* 16(2): 47.

BLEEKER, P. 1859. *Epicrium monochrous* Blkr. *Natuur. Tijd. Ned-Indië* 16(2): 188.

BLEEKER, P. 1859. Verslag omtrent vischsoorten en reptiliën van Ngawi, verzameld door J. T. van Bloemen Waanders en F. W. Sythoff. *Tijd. Ned-Indië* 16(2): 357–360.

BLEEKER, P. 1859. Reptiliën van Fort Kock, verzemeld door E. W. A. Ludeking. *Tijd. Ned-Indië* 16(2): 388.

BLEEKER, P. 1859. Verslag van reptiliën en visschen van Westlijke Borneo, aangeboden door J. H. A. B. Sonnemann Rebenitsch. *Natuur. Tijd. Ned-Indië* 16(2): 433–441.

BLEEKER, P. 1859. Over vischoorten en reptiliën van Bintang en Siak, aageboden door den heer E. F. Meijer. *Natuur. Tijd. Ned-Indië* 20(1): 86–88.

BLEEKER, P. 1859. Reptiliën van Sintang. *Natuur. Tijd. Ned-Indië* 20(6): 200–201.

BLEEKER, P. 1859. *Sphargis coriacea*, gevangen bij Cheribon. *Natuur. Tijd. Ned-Indië* 20(6): 204

BLEEKER, P. 1860. Reptiliën uit de omstreken van Tandjong. *Natuur. Tijd. Ned-Indië* 20(6): 220–221.

BLEEKER, P. 1860. Reptiliën van Agam. *Natur. Tijdsch. v. Nederl. Indië* 20(6): 325–329.

BLEEKER, P. 1860. Slangen van Rau (Sumatra's westkust), aangeboden door J. A. James. *Natuur. Tijd. Ned-Indië* 20(6): 416.

BLEEKER, P. 1860. Reptiliën van Bankalis. *Natuur. Tijd. Ned-Indië* 20(6): 416.

BLEEKER, P. 1860. Bestuursvergadering, gehouden den 12n January 1860, ten huize van den Heer Bleeker. 7. Brief van het lid, den heer P. L. Bloemen Waanders, van Banjoewangi, aanbiedende een slang van Bali op spiritus, welke bij de islanders aldaar als een vliegende bekend staat. *Natuur. Tijd. Ned-Indië* 20(6): 470–471.

BLEEKER, P. 1860. Over de reptiliën-fauna van Sumatra. *Natuur. Tijd. Ned-Indië* 21(1): 284–298.

BLEEKER, P. 1860. Reptiliën van Buitenzorg. *Natuur. Tijd. Ned-Indië* 21(1): 331.

BLEEKER, P. 1860. Soorten van reptiliën van het eiland Banka. *Natuur. Tijd. Ned-Indië* 21(1): 331–334.

BLEEKER, P. 1860. Reptiliën van Japara. *Natuur. Tijd. Ned-Indië* 21(1): 333.

BLEEKER, P. 1860. Reptiliën uit de omstreken van Fort de Kock. *Natuur. Tijd. Ned-Indië* 21(1): 437–438.

BLEEKER, P. 1860. Over de reptiliën-fauna van van (sic) Ceram. *Natuur. Tijd. Ned-Indië* 22(2): 35–38.

BLEEKER, P. 1860. Over de reptiliën-fauna van Amboina. *Natuur. Tijd. Ned-Indië* 22(2): 39–43.

BLEEKER, P. 1860. Thans bekende reptiliën van Celebes. *Natuur. Tijd. Ned-Indië* 22(2): 83–85.

BLEEKER, P. 1860. Thans bekende reptiliën van Timor. *Natuur. Tijd. Ned-Indië* 22(2): 86–88.

BLEEKER, P. 1860. Over de reptiliën-fauna van Sumatra. *Natuur. Tijd. Ned-Indië* 22(2): 284–298.

BLEEKER, P. 1865. *Pareas laevis* van Batavia. *Natuur. Tijd. Ned-Indië* 38(3): 418.

BLOUCH, R. A., T. SINAGA & P. SUMARTO. 1981. Leatherback turtles return to Sukamade beach. *Tigerpaper* 8(1): 124–125.

BLYTH, E. 1846. Notes on the fauna of the Nicobar Islands. *J. Asiatic. Soc. Bengal* 15: 367–379.

BLYTH, E. 1859. Report of Curator, Zoological Department, for February to May Meeting, 1859. *J. Asiatic Soc. Bengal* 28(3): 671–679.

BLYTH, E. 1863. The zoology of Andaman Islands. *In*: Adventures and researches among the Andaman Islanders. pp: 345–367. F. J. Mouat (author). *J. Asiatic Soc.* Bengal 32: i–viii + 1–367. Reprinted 1979, 1989, as "The Andaman Islanders" by Mittal Publications, New Delhi.

BODDAERT, P. 1770. III. Over de Kraakbeenige schildpad. De Testudine cartilaginea. Kornelis van Tongerlo, Amsterdam. 40 pp.

BÖHME, W. 1982. Über Schmetterlingsagamen, *Leiolepis b. belliana* (Gray, 1827) der Malayischen Halbinsel und ihre parthenogenetischen Linien (Sauria: Uromastycidae). *Zool. Jb. Syst.* 109: 157–169.

BÖHME, W. 1989. Rediscovery of the Sumatran agamid lizard *Harpesaurus beccarii* Doria 1888, with the first note on a live specimen. *Trop. Zool.* 2: 31–35.

BOETTGER, O. 1873. *Calamaria iris* n. sp., neue Schlange von Sumatra. *Ber. Senckberg. Nat. Ges. 1892:* 65–164.

BOETTGER, O. 1883. Herpetologische Notizen: II. Listen von Reptilien und Amphibien der niederländisch-indischen Insel Bangka, der siamesischen Insel Salaga u. von Atschin in Nord-Sumatra. *Ber.-Senckenberg. Naturf. Ges.* 1883: 147–157.

BOETTGER, O. 1886. Über die aufgestellten Reptilien von Deli N. Sumatra. *Ber.-Senckenberg. Naturf. Ges.* 1886: 81–86.

BOETTGER, O. 1886. Aufzählung der von den Philippinen bekannten Reptilien und Batrachier. *Ber. Senckberg. Nat. Ges.* 1886: 91–134.

BOETTGER, O. 1887. Herpetologische Notizen: I. Listen von Reptilien und Batrachiern aus niederländisch-indischen und von der Insel Salanga. *Ber.-Senckenberg. Naturf. Ges.* 1887: 37–55.

BOETTGER, O. 1888. Aufzählung einiger neu erworbener Reptilien und Batrachier aus Ost-Asien. *Ber. Offenbach Ges.* 26–28: 187–190.

BOETTGER, O. 1892. Drei neue Colubriforme Schlangen. *Zool. Anz.* 15: 418–420.

BOETTGER, O. 1892. Liste von Kriechtieren und Lurchen aus dem tropischen Asien u. ous Papuasien. *Ber. Offenbach Ges.* 29/32: 65–164.

BOETTGER, O. 1893. Neue Reptilien und Batrachier aus West-Java. *Zool. Anz.* 16: 334–340.

BOETTGER, O. 1895. Liste der Reptilien und Batrachier der Inseln Halmahera nach den Sammlungen Prof. W. Kukenthal's. *Zool. Anz.* 18(471): 116–121.

BOETTGER, O. 1895. Liste der Reptilien und Batrachier der Inseln Halmahera nach den Sammlungen Prof. W. Kukenthal's. *Zool. Anz.* 18(472): 129–138.

BOETTGER, O. 1898. Katalog der Reptilien-Sammlung im Museum der Senckenbergischen Naturfoschenden Gesslschaft in Frankfurt am Main. II Teil (Schlangen). Senckenbergische naturf. Gesellschaft, Frantfurt am Main. ix + 160 pp.

BOETTGER, O. 1901. Die Reptilien und Batrachier. Ergebnisse einer zoologischen Forschungsreise in den Molukken und Borneo. *In*: W. Kükenthal.

(Ed.). Zweiter Teil. Wissenschaftliche Reiseergebnisse. Band III. *Abh. Sencken-berg. Naturf. Ges.* 25: 321–402.

BOGERT, C. M. 1945. *Hamadryas* preoccupied for the king cobra. *Copeia* 1945(1): 47.

BOIE, F. 1826. Generalübersicht der Familien und Gattung der Ophidier. *Isis von Oken* 19(10): columns 981–982.

BOIE, F. 1827. Bemerkungen über Merrem's Versuch eines Systems der Amphibien. 1te Lieferung: Ophidier. *Isis (von Oken)* 20(6): columns 508–566.

BOIE, H. Unpublished, n.d. Erpétologie de Java. Manuscript deposited at Rijksmuseum van Natuurlijke Historie, Leiden.

BOIE, H. 1826. Merkmale einiger japanischen Lurche. *Isis von Oken* 19(2): columns 203–216.

BOIE, H. 1826. Notice sur l'Erpetologie de l'Ile de Java. *Bull. Sci. Nat. Geol. Paris* 9: 233–240.

BOIE, H. 1831. Briefe von Heinrich Boie geschreiben aus Ostindien und auf der Reise dahin. Neues Staatsbürg. *Mag. Holstein* 1(1): 126–218, 1(2): 440–500.

BOLT, G. H. & L. R. SCHULTZ. 1973. Eenige ervaringen met de verzorging van de Indonesische skink, *Mabuya multifasciata* Kuhl. *Lacerta* 31(5): 71–79.

BONAPARTE, C. L. 1845. Specchio generale dei sistemi erpetologico ed anfibiologico. *Atti Sci. Ital. (Milano)* 6: 376–378.

BOOTH, W. E., K.-M. WONG, A. S. KAMARIAH, S. C. CHOY & I. DAS. 1997. A survey of the flora and fauna of Pulau Punyit, Brunei Darussalam. *Sandakania* 9: 55–66.

BORG, J. P. & H. T. BORG. 1981. De tokkeh (*Gekko gecko*), een assertief reptiel. *Lacerta* 39(6 & 7): 68–70.

BOSCHMA, H. 1922. Über den Trichterapparat der Larven von *Megalophrys montana* Kuhl. *Bijdr. tot Dierk* 22: 9–12.

BOULENGER, G. A. 1882. Catalogue of the Batrachia Salientia s. Ecaudata in the collection of the British Museum. Taylor & Francis, London. xvi + 503 pp + PL. I–XXX.

BOULENGER, G. A. 1885. Catalogue of lizards in the British Museum (Natural History). Second edition. Vol. 1. Geckonidae [sic], Eublepharidae, Uroplatidae, Pygopodidae, Agamidae. British Museum (Natural History), London. xii + 436 pp + Pl. I–XXXII.

BOULENGER, G. A. 1885. Catalogue of lizards in the British Musum (Natural History). Second edition. Volume II. Iguanidae, Xenosauridae, Zonuridae, Anguidae, Anniellidae, Helodermatidae, Varanidae, Xantusiidae, Teiidae, Amphiesbaenidae. British Museum (Natural History), London. xiii + 497 pp + Pl. I–XXIV.

BOULENGER, G. A. 1885. A list of reptiles and batrachians from the island of Nias. *Ann. & Mag. nat. Hist. Ser. 5* 16: 388–389.

BOULENGER, G. A. 1887. Catalogue of lizards in the British Museum (Natural History). Second edition. Volume III. Lacertidae, Gerrhosauridae,

Scincidae, Anelytropidae, Dibamidae, Chamaeleontidae. British Museum (Natural History), London. xii + 575 pp + Pl. I–XL.

BOULENGER, G. A. 1887. Description of a new snake, of the genus *Calamaria*, from Borneo. *Ann. & Mag. nat. Hist. Ser. 5* 19: 169–170.

BOULENGER, G. A. 1887. On new reptiles and batrachians from North Borneo. *Ann. & Mag. nat. Hist. Ser. 5* 20: 95–97.

BOULENGER, G. A. 1887. Note on some reptiles from Sumatra described by Bleeker in 1860. *Ann. & Mag. nat. Hist. ser 5* 20: 152.

BOULENGER, G. A. 1888. Note on the classification of the Ranidae. *Proc. Zool. Soc. London* 1888(2): 204–206.

BOULENGER, G. A. 1889. Note on *Python curtus. Proc. Zool. Soc. London* 1889: 432–433; Pl. 45.

BOULENGER, G. A. 1890. List of the reptiles, batrachians and freshwater fishes collected by Prof. Moesch and Mr. Iversen in the district of Deli, Sumatra. *Proc. Zool. Soc. London* 1890: 31–40.

BOULENGER, G. A. 1890. Second report on additions to the batrachian collection in the Natural History Museum. *Proc. Zool. Soc. London* 1890: 323–328.

BOULENGER, G. A. 1891. Descriptions of new Oriental reptiles and batrachians. *Ann. & Mag. nat. Hist. Ser. 6* 7: 279–283.

BOULENGER, G. A. 1891. On new or little known Indian and Malayan reptiles and batrachians. *Ann. & Mag. nat. Hist. Ser. 6* 8: 288–292.

BOULENGER, G. A. 1891. Remarks on the herpetological fauna of Mount Kina Baloo, North Borneo. *Ann. & Mag. Nat. Hist. Ser. 6* 7: 341–345.

BOULENGER, G. A. 1892. An account of the reptiles and batrachians collected by Mr. C. Hose on Mt. Dulit, Borneo. *Proc. Zool. Soc. London* 1892: 505–508.

BOULENGER, G. A. 1893. Descriptions of new reptiles and batrachians obtained in Borneo by Mr. A. Everett and Mr. C. Hose. *Proc. Zool. Soc. London* 1893: 522–528.

BOULENGER, G. A. 1893. Catalogue of the snakes in the British Museum (Natural History). Vol. I. Containing the families Typhlopidae, Glauconiidae, Boidae, Ilysiidae, Uropeltidae, Xenopeltidae and Colubridae aglyphae. British Museum (Natural History), London. xiii + 448 pp + Pl. I–XXVIII.

BOULENGER, G. A. 1894. A list of the reptiles and batrachians collected by Dr. E. Modigliani on Sereinu (Sipora), Mentawei Islands. *Ann. Mus. Civ. Genova Ser. 2* 14: 613–618.

BOULENGER, G. A. 1894. Catalogue of the snakes in the British Museum (Natural History). Vol. II. Containing the conclusion of the Colubridae aglyphae. British Museum (Natural History), London. ix + 382 pp + Pl. I–XX.

BOULENGER, G. A. 1894. On the herpetological fauna of Palawan and Balabac. *Ann. & Mag. nat. Hist. Ser. 6* 14: 81–90.

BOULENGER, G. A. 1895. Descriptions of four new batrachians discovered by Mr. Charles Hose in Borneo. *Ann. & Mag. nat. Hist. Ser. 6* 16: 169–171.

BOULENGER, G. A. "1895" 1896. Descriptions of new reptiles and batrachians collected in Celebes by Drs. P. and F. Sarasin. *Ann. & Mag. nat. Hist. Ser. 6* 17: 393–395.

BOULENGER, G. A. "1895" 1896. Descriptions of new batrachians in the British Museum. *Ann. & Mag. nat. Hist. Ser. 6* 17: 401–406.

BOULENGER, G. A. 1896. Catalogue of the snakes in the British Musum (Natural History). Volume III., containing the Colubridae (Opisthoglyphae and Proteroglyphae), Amblycephalidae, and Viperidae. British Museum, London. xiv + 727 pp + Pl. I–XXV.

BOULENGER, G. A. 1896. Descriptions of new reptiles and batrachians collected in Celebes by Drs. P. & F. Sarasin. *Ann. & Mag. nat. Hist. Ser. 6* 17: 393–395.

BOULENGER, G. A. 1896. Descriptions of two new batrachians obtained by Mr. A. Everett on Mount Kina Balu, North Borneo. *Ann. & Mag. nat. Hist. Ser. 6* 17: 449–450.

BOULENGER, G. A. 1896. Descriptions of new reptiles and batrachians obtained by Mr. Alfred Everett in Celebes and Jampea. *Ann. & Mag. nat. Hist. Ser. 6* 18: 62–64.

BOULENGER, G. A. 1897. Descriptions of new Malay frogs. *Ann. & Mag. nat. Hist. Ser. 6* 19: 106–108.

BOULENGER, G. A. 1897. A list of reptiles and batrachians collected by Mr. Alfred Everett in Lombok, Flores, Sumba and Savu, with description of new species. *Ann. & Mag. nat. Hist. Ser. 6* 19: 503–509.

BOULENGER, G. A. 1897. A catalogue of the reptiles and batrachians of the Celebes, with special reference to the collections made by Drs. P. and F. Sarasin, 1893–1896. *Proc. Zool. Soc. London* 13: 193–237; Pl. 7–16.

BOULENGER, G. A. 1897. Description of a new genus and species of tortoise from Borneo. *Ann. & Mag. nat. Hist. Ser. 6* 19: 468–469.

BOULENGER, G. A. 1898. Description of a new sea-snake from Borneo. *Proc. Zool. Soc. London* 1898: 106–107.

BOULENGER, G. A. 1898. Description of a new genus of aglyphous colubrine snakes from Sumatra. *Ann. & Mag. nat. Hist. Ser. 7* 2: 73–74.

BOULENGER, G. A. 1899. Descriptions of three new reptiles and a new batrachian from Mount Kina Balu, North Borneo. *Ann. & Mag. nat. Hist. Ser. 7* 4: 451–454.

BOULENGER, G. A. 1900. Descriptions of new reptiles and batrachians from Borneo. *Proc. Zool. Soc. London* 1900: 182–187; Pl. XIV–XVII.

BOULENGER, G. A. 1904. Description of a new genus of frogs of the family Dyscophidae, and list of the genera and species of that family. *Ann. & Mag. nat. Hist. Ser. 7* 13: 42–44.

BOULENGER, G. A. 1908. A revision of the Oriental pelobatid batrachians (genus *Megalophrys*). *Proc. Zool. Soc. London* 1908: 407–430; Pl. 22–25.

BOULENGER, G. A. 1912. A vertebrate fauna of the Malay Peninsula from the Isthmus of Kra to Singapore including the adjacent islands. Reptilia and Batrachia. Taylor and Francis, London. xiii + 294 pp.

BOULENGER, G. A. 1918. Remarks on the batrachian genera *Cornufer*, Tschudi, *Platymantis* Gnthr., *Simomantis* g.n., and *Staurois*, Cope. *Ann. & Mag. nat. Hist. Ser. 9* 1: 372–375.

BOULENGER, G. A. 1920. Monograph of the Lacertidae. Volume 1. British Museum (Natural History), London. x + 352 pp.

BOULENGER, G. A. 1920. A monograph of the south Asia, Papuan, Melanesian, and Australian frogs of the genus *Rana*. *Rec. Indian Mus.* 20: 1–226.

BOULENGER, G. A. 1920. Reptiles and batrachians collected in Korinchi, West Sumatra, by Messr. H. C. Robinson & C. Boden Kloss. *J. Fed. Malay States Mus.* 8: 285–306; Pl. VIII.

BOULENGER, G. A. 1920. Descriptions of a new gecko and a new snake from Sumatra. *Ann. & Mag. nat. Hist. Ser. 9* 5: 281–283.

BOULENGER, G. A. 1921. Monograph of the Lacertidae. Volume 2. British Museum (Natural History), London. viii + 451 pp.

BOUR, R. & A. DUBOIS. 1983. Nomenclatural availability of *Testydi ciruacea* Vandelli, 1761: A case against a rigid application of the rules to old, well-known zoological works. *J. Herpetol.* 17(4): 356–361.

BOUR, R., A. DUBOIS & R. G. WEBB. 1995. Types of recent trionychid turtles in the Muséum national d'Histoire naturelle, Paris. *Dumerilia* 2: 73–92.

BRAUN, W. & C. BRAUN. 1990. Zur Haltung und Zucht von *Cosymbotes platyurus* (Schneider, 1792). *Herpetofauna* 12(67): 27–30.

BROEKMEIJER, J. G. Z. & J. HAGEMAN. 1859. Reptiliën van Ost-Java. *Natuur. Tijd. Ned-Indië* 16(2): 310–311.

BR(tm)ER, W. 1978. Rotschwanznatter *Goniosoma oxycephala*, ihre Pflege und Zucht. *Das Aquarium* 104: 79–81.

BR(tm)ER, W. 1981. Über die erfolgreiche Nachzucht einer F_2-Generation der Rotschwanznatter, *Gonyosoma oxycephalum* (Boie 1827). *Salamandra* 17(1/2): 86–87.

BRONGERSMA, L. D. 1928. Lizards from Pulu Berhala. *Misc. Zool. Sumatra* (26): 1–3.

BRONGERSMA, L. D. 1929. A list of reptiles from Java. *In*: On the zoogeography of Java. K. W. Dammerman (Ed.). *Treubia* 11: 64–68.

BRONGERSMA, L. D. 1930. Notes on the list of reptiles of Java. *Treubia* 12(3–4): 299–303.

BRONGERSMA, L. D. 1930. Abnormal coloration of *Xenopeltis unicolor* Reinw (1827). *Copeia* 1930(3): 87.

BRONGERSMA, L. D. 1931. Résultats scientifiques du voyage aux Indes Orientales Néerlandaises du Prince et de la Princesse Léopold de Belgique, Reptilia. *Verh. Koninkl. Natuurhistorisch Museum, Brussels* 5(2): 1–39; Pl. I–IV.

BRONGERSMA, L. D. 1932. Some notes on the genus *Hemiphyllodactylus* Bleeker. *Zool. Med.* 14: 1–13; 1 map.

BRONGERSMA, L. D. 1932. Über die Eiablage und die Eier von *Varanus komodoensis* Ouwens. *Zool. Gart., Leipzig (n.s.)* 5(1/3): 45–48.

BRONGERSMA, L. D. 1933. Additions to the reptile-fauna of the Batu Islands. *Misc. Zool. Sumatra* (72): 1–2.

BRONGERSMA, L. D. 1933. On a young *Gymnodactylus marmoratus* D. B. from Pulu Berhala. *Misc. Zool. Sumatra* (73): 1–3.

BRONGERSMA, L. D. 1933. The herpetological fauna of Pulu Weh. Herpetological notes V. *Zool. Med.* 16: 12–17.

BRONGERSMA, L. D. 1933. Bemerkungen über einige angeblich aus dem Indo-Australischen Archipel stammende Reptilien. *Mitt. Zool. Mus. Berlin* 18(3): 319–321.

BRONGERSMA, L. D. 1934. Contributions to Indo-Australian herpetology. *Zool. Med.* 17: 161–251; Pls. 1–2.

BRONGERSMA, L. D. 1935. Notes on the newly described genus *Cacophryne* Davis. *Zool. Med.* 18: 257–259.

BRONGERSMA, L. D. 1935. Ein "neuer" Javanischer Fundort von *Phrynoglossus laevis laevis* (Gthr). *Zool. Med.* 18: 265–266.

BRONGERSMA, L. D. 1937. On a small collection of Amphibia from central east Borneo. *Zool. Med.* 20: 6–9; 1 pl.

BRONGERSMA, L. D. 1943. On two *Rhacophorus* species mentioned by Kuhl & Van Hasselt. *Arch. Néerl. Zool. ser. 4e* 6: 341–346.

BRONGERSMA, L. D. 1945. Notes on the list of reptiles of Java. *Treubia* 12: 299–303.

BRONGERSMA, L. D. 1945. On the arrangement of the scales on the dorsal surface of the digits in *Lygosoma* and allied genera. *Zool. Med.* 24: 153–158.

BRONGERSMA, L. D. 1947. On the identity of *Maticora intermedia* Westermann and *Dipsadoides decipiens* Annandale. *Proc. Koninkl. Nederl. Akad. Wet. (C)* 50(4): 419–425.

BRONGERSMA, L. D. 1947. On the subspecies of *Python curtus* Schlegel occurring in Sumatra. *Proc. Koninkl. Nederl. Akad. Wet. (C)* 50(6): 666–671.

BRONGERSMA, L. D. 1947. Herpetologisch onderzoek van den Indischen Archipel. Handel XXX. *Ned. Nat. Geneesk. Congr.:* 132–153.

BRONGERSMA, L. D. 1948. Frogs and snakes from the island of Morotai (Moluccas). *Zool. Med.* 20: 306–310.

BRONGERSMA, L. D. 1948. Lizards from the island of Morotai (Moluccas). *Proc. Koninkl. Ned. Akad. Wet. Ser. C* 51: 486–495.

BRONGERSMA, L. D. 1948. Notes on *Maticora bivirgata* (Boie) and *Bungarus flaviceps* Reinh. *Zool. Med.* 30: 1–29.

BRONGERSMA, L. D. 1951. De arteria pulmonalis bij de Boidae en bij *Xenopeltis* (Serpentes). *Ned. Tijdschr. Geneesk.* 95(34): 2490–2491.

BRONGERSMA, L. D. 1952. Notes upon the arteries of the lungs in *Python reticulatus* (Schn.). *Proc. Koninkl. Ned. Akad. Wet. Ser. C* 55: 49–61.

BRONGERSMA, L. D. 1952. On the tracheal lung and lung in *Acrochordus* and some other snakes. *Arch. Néerl. Zool.* 9: 561–562.

BRONGERSMA, L. D. 1954. *Gymnodactylus marmoratus* (Gray). *Versl. gew. Verg. Afd. Natuur. Koninkl. Ned. Akad. Wet. Ser. C* 61(10): 172–175.

BRONGERSMA, L. D. 1956. On two species of boid snakes from the Lesser Sunda Islands. *Proc. Koninkl. Nederl. Akad. Wet. Ser. C* 59: 290–300.

BRONGERSMA, L. D. 1956. The palato-maxillary arch in some Asiatic Dipsadinae (Serpentes). *Proc. Konink. Ned. Akad. Wet. Ser. C* 59: 439–446.

BRONGERSMA, L. D. 1956. Dipsadinae can de Indische Archipel. *Versl. gew. Verg. Afd. Natuur. Koninkl. Ned. Akad. Wet. Ser. C* 65: 50–52.

BRONGERSMA, L. D. 1958. Some features of the Dipsadinae and Pareinae (Serpentes, Colubridae). *Proc. Konink. Ned. Akad. Wet. Ser. C* 61: 7–12.

BRONGERSMA, L. D. 1958. Note on *Vipera russelii* (Shaw). *Zool. Med.* 36: 55–76; pl. 1–3.

BRONGERSMA, L. D. 1958. *Vipera russelii* in the Indische Archipel. *Versl. gew. Verg. Afd. Natuur. Koninkl. Ned. Akad. Wet. Ser. C* 67: 88–91.

BRONGERSMA, L. D. 1958. On an extinct species of *Varanus* (Reptilia, Sauria) from the island of Flores. *Zool. Med.* 36: 113–125; Pl. 4–8.

BRONGERSMA, L. D. 1972. On the "Histoire Naturelle des Serpents" by De la Cépède 1789 and 1790, with a request to reject this work as a whole, with proposal to place seven names of snakes, being nomina oblita, on the Official Index of Rejected and Invalid names in Zoology, and to place three names of snakes on the Official List of Species-Names in Zoology (Class Reptilia). *Bull. Zool. Nom.* 29(1): 44–61.

BRONGERSMA, L. D. & W. HELLE. 1951. Notes on Indo-Australian snakes, I. *Proc. Konink. Ned. Akad. Wet. Ser. C* 54: 1–8.

BRONGERSMA, L. D., R. F. INGER & H. MARX. 1966. Proposed use of the plenary powers to conserve the generic name *Calamaria* Boie, 1827, and the specific name *Calamaria linnaei* Schlegel, 1827 (Reptilia, Serpentes). *Bull. Zool. Nom.* 22(5/6): 303–313.

BRONGERSMA, L. D. & C. WEHLBURG. 1933. Notes on some snakes from Sumatra. *Misc. Zool. Sumatra* 79: 1–6; 1 pl.

BROUGHTON, M. (LADY). 1936. A modern dragon hunt on Komodo. *Natl. Geogr.* 70(32): 321–331.

BROWN, A. E. 1902. A collection of reptiles and batrachians from Borneo and the Loo Choo Islands. *Proc. Acad. Nat. Sci. Philadelphia* 54: 175–186.

BROWN, A. E. 1902. A list of reptiles and batrachians in the Harrison-Hiller collection from Sumatra. *Proc. Acad. nat. Sci. Philadelphia* 1902: 693–695.

BROWN, M. W. 1986. The fierce and ugly Komodo Dragon fights on. *New York Times* 24 June, 1986.

BROWN, W. C. 1977. Lizards of the genus *Lepidodactylus* (Gekkonidae) from the Indo-Australian Archipelago and the islands of the Pacific, with descriptions of new species. *Proc. California Acad. Sci.* 41: 253–265.

BROWN, W. C. 1956. A revision of the genus *Brachymeles* (Scincidae), with descriptions of new species and subspecies. *Breviora* (54): 1–19.

BROWN, W. C. 1991. Lizards of the genus *Emoia* (Scincidae) with observations on their evolution and biogeography. *Mem. California Acad. Sci.* 15: 1–94.

BROWN, W. C. & A. C. ALCALA. 1956. A review of the Philippine lizards of the genus *Lygosoma* (*Leiolopisma*). *Occ. Pap. Nat. Hist. Mus. Stanford Univ.* (3): 1–10.

BROWN, W. C. & A. C. ALCALA. 1961. A new sphenomorphid lizard from Palawan Island, Philippines. *Occ. Pap. California Acad. Sci.* (32): 1–4.

BROWN, W. C. & A. C. ALCALA. 1962. A new lizard of the genus *Gekko* from the Philippine Islands. *Proc. Biol. Soc. Washington* 75: 67–70.

BROWN, W. C. & A. C. ALCALA. 1963. Additions to the leiolopismid lizards known from the Philippines, with descriptions of a new species and subspecies. *Proc. Biol. Soc. Washington* 76: 69–80.

BROWN, W. C. & A. C. ALCALA. 1970. The zoogeography of the herpetofauna of the Philippines Islands, a fringing archipelago. *Proc. Calif. Acad. Sci. ser. 4* 38: 105–130.

BROWN, W. C. & A. C. ALCALA. 1977. Diagnoses of species of *Lepidodactylus* (Gekkonidae) from the Indo-Australian Archipelago and the islands of the Pacific Basin, with descriptions of new species. *Proc. California Acad. Sci.* 41: 253–265.

BROWN, W. C. & A. C. ALCALA. 1978. Philippine lizards of the family Gekkonidae. Silliman University Natural Science Monograph Series No. 1, Dumaguete City. v + 146 pp + (1) errata sheet.

BROWN, W. C. & A. C. ALCALA. 1980. Philippine lizards of the family Scincidae. Silliman University Natural Science Monograph Series No. 2, Dumaguete City. xi + 264 pp + (1) errata sheet.

BRYANT, C. J. 1982. The red-tailed racer, *Gonyosoma* (*Elaphe*) *oxycephala*. *The Herpetile* 7(3): 33–34.

BURDEN, W. D. 1927. The quest for the dragon. *Nat. Hist.* 27: 3–18.

BURDEN, W. D. 1927. Dragon lizards of Komodo. Putnam Sons, New York. 221 pp.

BURDEN, W. D. 1927. Stalking the dragon lizard on the island of Komodo. *Nat. Geogr.* 52(2): 216–233.

BURDEN, W. D. 1928. Observations on the habits and distribution of *Varanus komodoensis* Ouwens. *American Mus. Novit.* (316): 1–10.

BURGER, W. L. 1971. Genera of pitvipers (Serpentes: Crotalidae). Ph. D. Dissertation, University of Kansas, Lawrence, Kansas. 186 pg.

BURGER, W. L. 1974. A case of mild envenomation by the mangrove snake, *Boiga dendrophila*. *The Snake* 6: 99–100.

BURGER, W. L. & T. NATSUNO. 1974. A new genus for the Arafura smooth snake and redefinitions of other seasnake genera. *The Snake* 6: 61–75.

BUSKIRK, J. R. 1989. A third specimen and neotype of *Heosemys leytensis* (Chelonia: Emididae). *Copeia* 1989(1): 224–227.

BUSONO, P. 1974. Facts about the *Varanus komodoensis* at the Gembiro Loka Zoo. *Zool. Gart., Jena* 44: 62–63.

BUTLER, A. L. 1905. The eggs and embryos of Schlegel's gavial (*Tomistoma schlegeli*, S. Müller). *J. Fed. Malay St. Mus.* 1(1): 1–2.

CANTOR, T. E. 1836. Sketch of an undescribed hooded serpent, with fangs and maxillar [sic] teeth. *Asiatic Res.* 19: 87–93, Pl. X–XII.

CANTOR, T. E. 1838. A notice of the *Hamadryas*, a genus of hooded serpents with poisonous fangs and maxillary teeth. *Proc. Zool. Soc. London* 1838: 72–75.

CANTOR, T. E. 1839. Spicegium serpentium indicorum. *Proc. Zool. Soc. London* 1839: 31–34; 49–55.

CANTOR, T. E. 1847. Catalogue of reptiles inhabiting the Malayan Peninsula and Islands. *J. Asiatic Soc. Bengal* 16: 897–952, 1026–1078.

CAPOCACCIA, L. 1961. Catalogo dei tipi di rettili del Museo Civico di Storia Naturale di Genova. *Ann. Mus. Civico Stor. Nat. Genova* 72: 86–111.

CARD, W. 1994. Notes on the natural history and husbandry of the temple pit viper (*Tropidolaemus wagleri*). *The Vivarium* 5(5): 22–25.

CARR, J. L. & J. W. BICKHAM. 1981. Sex chromosomes of the Asian pond turtle, *Siebenrockiella crassicollis* (Testudines: Emydidae). *Cytogenetics Cell Genetics, Basel* 31: 178–183.

CHANNING, A. 1989. A re-evaluation of the phylogeny of Old World treefrogs. *S. African J. Zool.* 24: 116–131.

CHAPMAN, P. 1984. Cave-frequenting vertebrates in the Gunung Mulu National Park, Sarawak. *Sarawak Mus. J.* 33(54): 101–105.

CHASEN, F. H. & N. SMEDLEY. 1927. A list of reptiles from Pulo Galang and other islands of the Rhio Archipelago. *J. Malay Br. Roy. Asiatic Soc.* 5(1–3): 351–355.

CHERCHI, M. A. 1954. Una nuova sottospecies di *Kaloula baleata* delle isole Andamane. *Doriana* 1(47): 1–4.

CHIN, L. 1975. Notes on marine turtles (*Chelonia mydas*). *Sarawak Mus. J.* 23: 259–265.

CHIN, L. 1976. Tom Harrisson and the green turtles of Sarawak. *Borneo Res. Bull.* 8: 2.

CHOUDHURY, B. C. & H. R. BUSTARD. 1979. Predation on natural nests of the saltwater crocodile (*Crocodylus porosus* Schneider) on North Andaman Island with notes on the crocodile population. *J. Bombay nat. Hist. Soc.* 75(2): 43–49.

CHRAPLIWY, P. S., H. M. SMITH & C. GRANT. 1961. Systematic status of the geckonid lizard genera *Gehyra, Peropus, Hoplodactylus* and *Naultinus*. *Herpetologica* 17(1): 5–12.

CHRISTIAN, A. & T. GARLAND, Jr. 1996. Scaling of limb proportions in monitor lizards (Squamata: Varanidae). *J. Herpetol.* 30(2): 219–230.

CHRISTENSEN-DALSGAARD, J. T. LUDWIG & P. M. NARINS. 1997. Calling behavior of the south-east Asian rhacophorid frog *Polypedates leucomystax*. *In:* Herpetology '97. Abstracts of the Third World Congress of Herpetology. 2–10 August, 1997. Z. Rocek & S. Hart (Eds). pp: 41. Third World Congress of Herpetology, Prague.

CHUANG, L.-C., H.-M. YU, C.-P. CHEN, T.-H. HUANG, S.-H. WU & K.-T. WANG. 1996. Determination of three-dimensional solution structure of waglerin I, a toxin from *Trimeresurus wagleri*, using 2D-NMR and molecular dynamics simulation. *Biochim. Biophy. Acta. Protein Structure & Mol. Enzymol.* 1292(1): 145–155.

CHUNG, M.-C.-M., G. PONNUDURAI, M. KATAOKA, S. SHIMIZU & N.-T. TAN. 1996. Structural studies of a major hemorrhagin (rhodostoxin) from the venom of *Calloselasma rhodostoma* (Malayan pit viper) *Arch. Biochem. & Biophy.* 325(2): 199–208.

CHURCH, G. 1959. Size variation in *Bufo melanostictus* from Java and Bali. *Treubia* 25(1): 113–126.

CHURCH, G. 1960. Annual and lunar periodicity in the sexual cycle of the Javanese toad, *Bufo melanostictus* Schneider. *Zool.* 44: 181–188.

CHURCH, G. 1960. The invasion of Bali by *Bufo melanostictus. Herpetogica* 16(1): 15–22.

CHURCH, G. 1960. The effects of seasonal and lunar changes on the breeding pattern of the edible Javanese frog *Rana cancrivora* Gravenhorst. *Treubia* 25(2): 215–233.

CHURCH, G. 1960. A comparison of a Javanese and a Balinese population of *Bufo biporcatus* with a population from Lombok. *Herpetologica* 16(1): 23–28.

CHURCH, G. 1962. The reproductive cycles of the Javanese house gecko, *Cosymbotus platyurua, Hemidactylus frenatus* and *Peropus mutilatus. Copeia* 1962(2): 262–269.

CHURCH, G. 1963. The variation of dorsal pattern in *Rhacophorus leucomystax. Copeia* 1963(2): 400–405.

CHURCH, G. & C. S. LIM. 1961. The distribution of three species of house geckos in Bandung (Java). *Herpetologica* 17(3): 119–201.

CIOFI, C., M. BRUFORD & I. R. SWINGLAND. 1997. Conservation biology and population genetics of the Komodo dragon *Varanus komodoensis. In*: Herpetology '97. Abstracts of the Third World Congress of Herpetology, 2–10 August, 1997. Z. Rocek & S. Hart (Eds). pp: 247. Third World Congress of Herpetology, Prague.

COCHRAN, D. M. 1926. A new pelobatid batrachian from Borneo. *J. Washington Acad. Sci.* 16: 446–447.

COCHRAN, D. M. 1933. Two new species of *Calamaria* from Borneo. *Proc. Biol. Soc. Washington.* 36: 91–92.

COHN, L. 1905. Schlangen aus Sumatra. *Zool. Anz.* 29(17): 540–548.

COLBERT, E. H. 1967. Adaptations for gliding in the lizard *Draco. American Novit.* (2283): 1–20.

COLE, C. J. & H. G. DOWLING. 1970. Chromosomes of the sunbeam snake, *Xenopeltis unicolor* Reinwardt (Reptilia: Xenopeltidae). *Herpetol. Rev.* 2(2): 35–36.

COLLENETTE, C. L. 1953. Strollers on the ceiling. *Malayan nat. J.* 8: 116.

COLLINS, P. 1956. My "Komodo dragon" circus. *Anim. Kingdom* 59(2): 34–42.

COOMANS-DE RUITER, L. 1955. De Indonesiche „Bushmasters" (*Trimeresurus*-soorten) I. *Lacerta* 14: 67–69.

COOMANS-DE RUITER, L. 1961. Aanvullende mededelingen over *Trimeresurus sumatranus. Lacerta* 19(5): 37–38.

COPE, E. 1860. Catalogue of Colubridae in the Museum of the Academy of Natural Sciences of Philadelphia. I. Calamarinae. *Proc. Acad. Nat. Sci. Philadelphia* 12: 74–79.

COPE, E. 1860. Catalogue of Colubridae in the Museum of the Academy of Natural Sciences of Philadelphia, with notes and descriptions of new species. Part 2. *Proc. Acad. Nat. Sci. Philadelphia* 12: 241–266.

COPE, E. "1860" 1861. Catalogue of the Colubridae in the Museum of the Academy of Natural Sciences of Philadelphia. Part 3. *Proc. Acad. Nat. Sci. Philadelphia* 12: 553–566.

COPE, E. 1895. The classification of the Ophidia. *Trans. American Phil. Soc.* 18(2): 186–219; Pl. 14–33.

COX, J. H., R. S. FRAZIER & R. A. MATURBONGS. 1993. Freshwater crocodiles of Kalimantan (Indonesian Borneo). *Copeia* 1993(2): 564–566.

COX, M. J. 1993. Some notes on the sunbeam snake, *Xenopeltis unicolor. Bull. Chicago Herpetol. Soc.* 28: 97.

CRANBROOK, EARL OF. 1981. The vertebrate faunas. *In*: Wallace's Line and Plate Tectonics. pp: 57–69. T. C. Whitmore (Ed.). Clarendon Press, Oxford.

CROMBIE, R. I. 1986. The status of the Nicobar toads *Bufo camortensis* Mansukhani & Sarkar, 1980 and *Bufo spinipes* Fitzinger *in* Steindachner, 1867. *J. Bombay nat. Hist. Soc.* 83: 226–229.

CROMBIE, R. I. In press. Key to the species of the scincid lizard genus *Mabuya* by David Horton with an introduction and a synopsis of the genus. *Smithsonian Herpetol. Inf. Serv.*

CRUMLY, C. R. 1982. A cladistic analysis of *Geochelone* using cranial osteology. *J. Herpetol.* 16(3): 215–234.

CRUMLY, C. R. 1984. A hypothesis for the relationship of land tortoise genera (family Testudinidae). Studia Geologica Salmanticensia. Vol. Especial I. *Studia Palaeochelonologica* 1: 115–124.

CRUMLY, C. R. 1988. A nomenclatural history of tortoises (family Testudinidae). *Smithsonian Herpetol. Inf. Serv.* 75: 1–17.

CUNDALL, D. & D. A. ROSSMAN. 1993. Cephalic anatomy of the rare Indonesian snake *Anomochilus weberi. Zool. J. Linn. Soc.* 109: 235–273.

CUNDALL, D., V. WALLACH & D. A. ROSSMAN. 1993. The systematic relationships of the snake genus *Anomochilus. Zool. J. Linn. Soc.* 109: 275–299.

CUVIER, G. L. C. F. D. "1817" 1816. Le règne animal distribué d'après son organisation. Tome II, contenant les Reptiles, les Poissons, les Mollusques et les Annélides. Déterville, Paris. xviii + 532 pp. (Reprinted 1969 Culture et Civilisation, Bruxelles.)

DALTRY, J. C. 1995. The evolutionary biology of the Malayan pit viper, *Calloselasma rhodostoma*: a study of the causes of intraspecific variation. Ph. D. Dissertation, University of Aberdeen, Aberdeen. x + 282 pg.

DALTRY, J. C., G. PONNUDURAI, C. K. SHIN, N.-H. TAN, R. S. THROPE & W. WÜSTER. 1996. Electrophoretic profiles and biological activities: intraspecific variation in the venom of the Malayan pit viper (*Calloselasma rhodostoma*). *Toxicon* 34(1): 67–79.

DALTRY, J. C., W. WÜSTER & R. S. THORPE. 1996. Diet and snake venom evolution. *Nature, London* 379: 537–540.

DALTRY, J. C., W. WÜSTER & R. S. THORPE. "1997" 1996. The role of ecology in determining venom variation in the Malayan pitviper, *Calloselasma rhodostoma*. *In*: Venomous snakes: ecology, evolution and snakebite. pp: 155–171. R. S. Thorpe, W. Wüster & A. Malhotra (Eds.). Symp. Zool. Soc. London (70). The Zoological Society of London/Clarendon Press, Oxford.

DAMMERMAN, K. W. 1923. The fauna of Krakatau, Verlaten Island and Sebesy. *Treubia* 3: 61–114.

DAMMERMAN, K. W. 1929. Tjibodas (Zoology). Fourth Pacific Science Congress, Java. Excursion C.3. The fauna of the Nature Reserve Tjibodas Gunung Gedeh. 30 pp.

DAMMERMAN, K. W. 1948. The fauna of Krakatau 1883–1933. *Verh. Koninkl. Ned. Akad. Wet. Sec. 2* 44: i–xii + 1–594 + 11 pl.

DANIELS, R. J. R. & P. V. DAVID. 1996. The herpetofauna of the Great Nicobar Island. *Cobra* 25: 1–4.

DAREVSKY, I. S.1962. Dragons are not necessarily cannibals. Tier 11: 31–33.

DAREVSKY, I. S. 1963. Zu Besuch bei den Komodo-Waranen. *Aquar. Terr.* 10(1): 74–80.

DAREVSKY, I. S. 1964. Die reptilien der Inseln Komodo, Padar and Rintja in Kleinen Sunda-Archipelago, Indonesia. *Senckenbergiana Biol.* 45(3/5): 563–576.

DAREVSKY, I. S. & W. AUFFENBERG. 1973. Growth records of wild, recaptured Komodo monitors (*Varanus komodoensis*). *HISS Newsl.* 1(2): 41–42.

DAREVSKY, I. S. & S. KADARSAN. 1964. On the biology of the giant Indonesian monitor lizard (*Varanus komodoensis*) Ouwens. *Zool. Zh.* 43(9): 1355–1360. Translated by Z. Knowles, *Smithsonian Herpetol. Inf. Serv.* (2): 1–6. 1965.

DAREVSKY, I. S. & I. M. STUSAK. 1968. A recapture and growth of a *Varanus komodoensis. Akad. Nauk.* 47(7): 1106–1107.

DAS, I. 1992. Eggs and hatchlings of some Bornean lizards. *Hamadryad* 17: 42–45.

DAS, I. 1992. Notes on the egg sizes and ticks of the yellow-lipped sea krait (*Laticauda colubrina*) on Pulau Punyit, Brunei Darussalam. *Hamadryad* 17: 45–46.

DAS, I. 1992. Frogs of Sabah [Book Review]. *Hamadryad* 17: 54–55.

DAS, I. 1992. The amphibian community at Batu Apoi, a lowland dipterocarp forest in Brunei Darussalam, north-western Borneo. *In:* Proceedings of the Asian Herpetological Meeting, 15–20 July, 1992. pp: 31–32. Huangshan, Anhui, China. Asian Herpetological Research Society, Huangshan.

DAS, I. 1993. The amphibians of Batu Apoi Forest Reserve. Universiti Brunei Darussalam, Bandar Seri Begawan. 6 pp.

DAS, I. 1993. The reptiles of Batu Apoi Forest Reserve. Universiti Brunei Darussalam, Bandar Seri Begawan. 6 pp.

DAS, I. 1993. Annandale's seasnake, *Kolpophis annandalei* (Laidlaw, 1901): A new record for Borneo (Reptilia: Serpentes: Hydrophiidae). *Raffles Bull. Zool.* 41(2): 359–361.

DAS, I. 1993. *Cnemaspis gordongekkoi*: A new gecko from Lombok, Indonesia, and the biogeography of the Oriental species of *Cnemaspis* (Squamata: Sauria: Gekkonidae). *Hamadryad* 18: 1–9.

DAS, I. 1993. Vernacular names of some southeast Asian amphibians and reptiles. *Sarawak Mus. J.* 34: 123–139.

DAS, I. 1994. Amphibian studies (pp: 279); Tree frogs (pp: 282); Amphibian diversity (pp: 283); Reptile diversity at Belalong (pp: 285); Turtles and tortoises (pp: 285–286); Household friends (pp: 287), Lizards of Belalong (pp: 288–289); Snakes of Belalong forest (pp: 290–291). *In:* Belalong: A tropical rainforest. The Earl of Cranbrook & David S. Edwards (Eds.). The Royal Geographical Society, London and Sun Tree Publishing Pte Ltd., Singapore.

DAS, I. 1994. Evaluating biodiversity: The Batu Apoi experience. *Ambio* 23(4/5): 238–242.

DAS, I. 1995. A new tree frog (genus *Polypedates*) from Great Nicobar, India (Anura: Rhacophoridae). *Hamadryad* 20: 13–20.

DAS, I. 1995. An illustrated key to the turtles of insular south-east Asia. *Hamadryad* 20: 27–32.

DAS, I. 1995 Amphibians and reptiles recorded from Batu Apoi, a lowland dipterocarp forest in Brunei Darussalam. *Raffles Bull. Zool.* 43(1): 157–180.

DAS, I. 1996. *Limnonectes shompenorum*, a new frog of the *Rana macrodon* (Anura: Ranidae) complex from Great Nicobar, India. *J. S. Asian nat. Hist.* 2(1): 60–67.

DAS, I. 1996. Geographic distribution: *Rana chalconota* (copper-cheeked frog). *Herpetol. Rev.* 27(1): 30.

DAS, I. 1996. Biogeography of the reptiles of south Asia. Krieger Publishing Company, Malabar, Florida. 16 colour plates + ix + 87 pp.

DAS, I. 1996. Spatio-temporal resource utilization by a Bornean rainforest herpetofauna: Preliminary results. *In:* Tropical Rainforest Research: Current Issues. pp: 315–323. D. S. Edwards, W. E. Booth & S. C. Choy (Eds.). Kluwer Academic Press, Dordrecht.

DAS, I. 1996. Status of knowledge of the biology and conservation of non-marine turtles of the Philippines. *In:* International Congress of Chelonian Conservation. Proceedings. pp: 81–83. SOPTOM (Ed.). Editions Soptom, Gonfaron.

DAS, I. 1996. Overview: The Asia-Pacific region. *In:* International Congress of Chelonian Conservation. Proceedings. pp: 98. SOPTOM (Ed.). Editions Soptom, Gonfaron.

DAS, I. 1996. The validity of *Dibamus nicobaricum* (Fitzinger *in* Steindachner, 1867)(Squamata: Sauria: Dibamidae). *Russian J. Herpetol.* 3(2): 157–162.

DAS, I. 1996. Snakes. *In:* Indonesian heritage. Volume 5. pp: 32–33. Wildlife. T. Whitten & J. Whitten (Eds). Editions Didier Millet/Archipelago Press, Singapore.

DAS, I. 1996. Lizards. *In:* Indonesian heritage. Volume 5. pp: 34–35. Wildlife. T. Whitten & J. Whitten (Eds). Editions Didier Millet/Archipelago Press, Singapore.

DAS, I. 1997. Rediscovery of *Lipinia macrotympanum* (Stoliczka, 1873) from the Nicobars Islands, India. *Asiatic Herpetol. Res.* 7: 23–26.

DAS, I. 1997. A reassessment of *Hardella isoclina* Dubois, 1908 (Testudines: Bataguridae) from the Trinil beds of the Javan Pleistocene. *Herpetol. J.* 7(2): 71–73.

DAS, I. 1997. A new species of *Cyrtodactylus* from the Nicobar Islands, India. *J. Herpetol.* 31(3): 375–382.

DAS, I. 1997. A new species of *Boiga* (Squamata: Serpentes: Colubridae) from the Nicobar Archipelago. *J. south Asian nat. Hist.* 3: 59–67.

DAS, I. 1997. Herpetological explorations in the Andaman and Nicobar Islands. *In:* Herpetology '97. Abstracts of the Third World Congress of Herpetology. 2–10 August, 1997. pp: 50. Z. Rocek & S. Hart (Eds). Third World Congress of Herpetology, Prague.

DAS, I. & H. V. ANDREWS. 1997. Bibliography of the herpetology of the Andaman and Nicobar Islands. *Hamadryad* 22(1): 69–72.

DAS, I. & K. CHANDRA. 1994. Two snakes new to Andaman and Nicobar Islands. *J. Andaman Sci. Assoc.* 10(1 & 2): 114–115.

DAS, I. & J. K. CHARLES. 1993. Amphibians and reptiles recorded from the Lambir Hills National Park, Sarawak, East Malaysia. *Hamadryad* 18: 17–23.

DAS, I. & J. K. CHARLES. 1993. A contribution to the herpetology of Bako National Park, Sarawak, East Malaysia. *Hamadryad* 18: 24–27.

DAS, I. & R. WHITAKER. 1996. A bibliography of the king cobra (*Ophiophagus hannah*). *Smithsonian Herpetol. Inf. Serv.* (108):1–24.

DAUDIN, F.-M. 1801 "1802". Histoire naturelle, génerale et particulière des reptiles. Vol. 2. F. Dufart, Paris. 432 pp.

DAUDIN, F.-M. 1802. Histoire naturelle, génerale et particulière des reptiles. Vol. 3. F. Dufart, Paris. 443 pp.

DAUDIN, F.-M. 1802. Histoire naturelle, génerale et particulière des reptiles. Vol. 4. F. Dufart, Paris. 395 pp.

DAUDIN, F.-M. 1803. Histoire naturelle, génerale et particulière des reptiles. Vol. 6. F. Dufart, Paris. 447 pp.

DAUDIN, F.-M. 1803. Histoire naturelle, génerale et particulière des reptiles. Vol. 7. F. Dufart, Paris. 436 pp.

DAUDIN, F.-M. 1803. Histoire naturelle, génerale et particulière des reptiles. Vol. 8. F. Dufart, Paris. 439 pp.

DAUDIN, F.-M. 1803. Caractère des vingt-trois genres qui composent l'order des Ophidiens. *Mag. Encyclopédique* (8e annee), 5(20): 433–438.

DAVENPORT, J. 1988. The turtle industry of Bali. *British Herpetol. Soc. Bull.* (25): 16–24.

DAVID, P. 1994. Liste des reptiles actuels du monde I. Chelonii. *Dumerilia* 1: 7–127.

DAVID, P. & G. VOGEL. 1996. The snakes of Sumatra: An annotated checklist and key with natural history notes. Edition Chimaira, Frankfurt am Main. 260 pp.

DAVIS, D. D. 1935. A new generic and family position for *Bufo borbonica*. *Field Mus. Nat. Hist., Zool. Ser.* 20: 87–92.

DAVIS, R., D. DARLING & A. DARLING. 1986. Ritualized combat in captive Dumeril's monitors, *Varanus dumerili*. *Herpetol. Rev.* 17(4): 85–86.

DAVIS, T. A. & R. ALTEVOGT. 1976. Giant turtles and robber crabs of the South Sentinel. *Yojana* 20(13): 75–79.

DAVISON, G. W. H. 1984. Foot-flagging in Bornean frogs. *Sarawak Mus. J.* 33: 177–178.

DAVISON, R. 1975. A fishing chichak. *Malayan Natural.* 1(4): 12.

DE ELERA, C. 1895. Catálogo sistemático de toda la fauna de Filipinas conocida hasta el presente, etc. Vertebrados 1. Santo Tomas College, Manila.

DE GRIJS, P. 1937. Eine neue Eidechse aus Nord-Borneo: *Calotes kinabaluensis. Zool. Anz.* 117: 136–138.

DE HAAS, C. P. J. 1934. Het jeugdkleed van *Dipsadomorphus cynodon*. *Trop. Natuur* 23: 154.

DE HAAS, C. P. J. 1934. Slangen vangsten. *Trop. Natuur* 23: 235–239.

DE HAAS, C. P. J. 1937. Eenige aasiteekeningen over *Haplopeltura boa* (Boie). *Trop. Natuur* 26: 101–104.

DE HAAS, C. P. J. 1938. *Naja bungarus. Trop. Natuur* 27: 37–38.

DE HAAS, C. P. J. 1941. Het genus *Natrix. Trop. Natuur* 30: 23–31.

DE HAAS, C. P. J. 1941. Some notes on the biology of snakes and their distribution in two districts of Java. *Treubia* 18(2): 327–375.

DE HAAS, C. P. J. 1950. Checklist of the snakes of the Indo-Australian Archipelago (Reptiles, Ophidia). *Treubia* 20: 511–625.

DEISSNER, F. H. 1859. Schildpadden van Banka. *Natuur. Tijd. Ned-Indië* 16(2): 316.

DEISSNER, F. H. 1860. Reptiliën uit de omstreken van Gombong. *Natuur. Tijd. Ned-Indië* 22(2): 85–86.

DE JONG, J. K. 1926. Fauna Buruana. Reptiles. *Treubia* 7(2): 85–96.

DE JONG, J. K. 1927. *Varanus komodoensis* Ouwens. *Ann. & Mag. nat. Hist. ser. 9* 19: 589–591.

DE JONG, J. K. 1928. Herpetologische Notizen (I). *Misc. Zool. Sumatrana* (32): 1–2.

DE JONG, J. K. 1928. Beiträge zur Kenntnis der Reptilienfauna von Niederländisch-ost-Indien. *Treubia* 10: 145–151.

DE JONG, J. K. 1930. Notes on some reptiles from the Dutch East Indies. *Treubia* 12(1): 115–119.

DE JONG, J. K. 1937. Ein en order over *Varanus komodoensis* Ouwens. *Natuurwet. Tijdschr. Ned.-Indië* 97(8): 173–208.

DE JONG, J. K. 1944. Newly hatched *Varanus komodoensis*. *Treubia* 18: 143–155.

DE JONG, J. K. & L. D. BRONGERSMA. 1927. Die Bewegungen im Schädel von *Varanus komodoensis*. *In*: Anatomische Notizen über *Varanus komodoensis* Ouwens. J. K. de Jong (Ed.). *Zool. Anz.* 70: 67–69.

DE LANG, D. & N. DE ROOIJ. 1910. Amphibien und Reptilien. *In*: Durch Zentral-Sumatra. II(III): 514–521. A. Maass (Ed.). B. Behr's Verlag, Berlin & Leipzig.

DE LISLE, H. F. 1996. The natural history of monitor lizards. Krieger Publishing Company, Malabar, Florida. xiii + 201 pp.

DELSMAN, H. C. 1926. Nog eens over krokodillen. *Trop. Natuur* 15: 55.

DELSMAN, H. C. 1927. Nog este over de varanen von Komodo. *Trop. Natuur* 16: 26–28.

DENSMORE, L. D. 1983. Biochemical and immunological systematics of the Order Crocodilia. *In*: Evolutionary biology. Vol. 16. pp: 397–465. M. K. Hecht, B. Wallace & G. H. Prance (Eds.). Plenum Press, New York.

DENSMORE, L. D. & H. C. DESSAUER. 1984. Low levels of protein divergence detected between *Gavialis* and *Tomistoma*: evidence for crocodilian monophyly? *Comp. Biochem. Physiol.* 77B: 715–720.

DENZER, W. 1994. Tree hole breeding in the toad *Pelophryne brevipes* (Peters, 1867). *Amphibia-Reptilia* 15: 224–226.

DERANIYAGALA, P. E P. 1944. Four new races of the 'kabaragoya' lizard *Varanus salvator*. *Spolia Zeylanica* 24: 59–62; Pl. 10–12.

DERANIYAGALA, P. E. P. 1947. The names of the water monitors of Ceylon, the Nicobars and Malaya. *Proc. Third Ann. Sess. Ceylon Assoc.* Sec. 2. Abstr. 12.

DERANIYAGALA, P. E. P. 1960. The water monitor of the Andaman Islands—a distinct subspecies. *Spolia Zeylanica* 29: 203–204.

DERANIYAGALA, P. E. P. 1960. The taxonomy of the cobras of south eastern Asia. *Spolia Zeylanica* 29: 205–232.

DERANIYAGALA, P. E. P. 1960. The water monitor of Andaman Islands—a distinct subspecies. *Spolia Zeylanica* 29: 203–224.

DERANIYAGALA, P. E. P. 1960. The taxonomy of the cobras of south eastern Asia. *Spolia Zeylanica* 29: 205–232.

DE ROOIJ, N. 1915. Reptiles. *In*: Die Insel Nias, zoologische Resultate. pp: 283–307. K. de Zwaan (Ed.). Martinus Nijhoff, Haak.

DE ROOIJ, N. 1915. The reptiles of the Indo-Australian Archipelago. I. Lacertilia, Chelonia, Emydosauria. E. J. Brill, Leiden. xiv + 384 pp.

DE ROOIJ, N. 1917. The reptiles of the Indo-Australian Archipelago. II. Ophidia. E. J. Brill, Leiden. xiv + 334 pp.

DE ROOIJ, N. 1922. Fauna Simalurensis. Reptilia. *Zool. Med.* 6(4): 217–238.

DE ROOIJ, N. 1922. List of reptiles from Krakatau, Verlaten Island and Sebesy. *Treubia* 3(1): 106.

DE SILVA, G. S. 1969. Turtle conservation in Sabah. *Sabah Soc. J.* 5: 6–26.

DE SILVA, G. S. 1969. Marine turtle conservation in Sabah. *Ann. Rep. Res. Board Forest Depart.* 1969: 124–135.

DE SILVA, G. S. 1971. Marine turtles in the State of Sabah, Malaysia. *IUCN n.s. Suppl. Pap.* 31: 47–52.

DE SILVA, G. S. 1978. Turtle notes (sightings of *Dermochelys* in Sabah waters). *Borneo Res. Bull.* 10(1): 23–24.

DE SILVA, G. S. 1980. The status of sea turtle populations in East Malaysia and the South China Sea. *In*: Biology and conservation of sea turtles. pp: 327–337. K. A. Bjorndal (Ed.). Smithsonian Institution Press, Washington, D.C.

DESPAIX, R. 1912. Sur trois collection de reptiles et de batraciens provenant de l'Archipel Malais. *Bull. Mus. Nat. Hist. nat. Paris* 18: 198–205.

DE VOOGD, C. N. A. 1939. Naar de Varanen van Komodo. 3. Jaren Indisch Natuur Leven, Nederlandsch Indosche Vareniging tat Natuurbescheming. Batavia. pp: 304–309.

DE VOOGD, C. N. A. 1950. Naar de Varanen van Komodo. *Trop. Natuur* 30: 128.

DE VOSJOLI, P. 1996. How to establish Reinwardt's gliding tree frogs *Rhacophorus reinwardtii. Vivarium* 7(5): 42–43.

DE WITTE, G. F. 1933. Liste des batraciens et des reptiles d'extrême-Orient et des Indes Orientales recuellis, en 1932, par S. A. R. le Prince Léopold de Belgique. *Bull. Mus. r. Hist. nat. Belg.* 9(24): 1–8.

DIAMOND, J. M. 1987. Natural selection: did Komodo dragons evolve to eat pygmy elephants? *Nature, London* 326: 832.

DIAMOND, J. M. 1992. The evolution of dragons in the jungles of Indonesia, those giant cold-blooded man-eating monters of yore have not only survived, they've climbed the top of the carnivore heap. *Discover* December, 1992: 72–80.

DJASMANI, H. & M. R. DJAMNARI. 1988. Monitor lizard populations in Kalimantan. Report to the CITES Secretariat. 10 pg.

DODD, C. K. 1988. Synopsis of the biological data on the loggerhead sea turtle *Caretta caretta* (Linnaeus 1758). *U.S. Fish & Wildl. Serv. Biol. Rep.* 88(14): 1–110.

DODD, C. K. 1990. *Caretta* Rafinesque Loggerhead turtles. *Cat. American Amphib.-Rept.* 482.1–482.2.

DODD, C. K. 1990. *Caretta caretta* (Linnaeus) Loggerhead sea turtle. *Cat. American Amphib.-Rept.* 483.1–483.7.

DOWLING, H. G. 1958. A taxonomic study of the ratsnakes VI. Validation of the genera *Gonyosoma* Wagler and *Elaphe* Fitzinger. *Copeia* 1958(1): 29–40.

DOWLING, H. G. 1993. The name of Russel's viper. *Amphibia-Reptilia* 14: 320.

DOWLING, H. G. & W. E. DUELLMAN. 1978. Systematic herpetology. A synopsis of families and higher categories. *HISS Publications* 7: i–vii + 1.1–118.3 + i–viii.

DOWLING, H. G. & F. W. GIBSON. 1970. Relationships of the Asian sunbeam snake, *Xenopeltis unicolor. Herpetol. Rev.* 2(3): 51–52.

DRING, J. C. 1983. Some frogs from Sarawak. *Amphibia-Reptilia* 4: 103–115.

DRING, J. C. 1983. Frogs of the genus *Leptobrachella* (Pelobatidae). *Amphibia-Reptilia* 4: 89–102.

DRING, J. C. 1987. Bornean treefrogs of the genus *Philautus* (Rhacophoridae). *Amphibia-Reptilia* 8: 19–47.

DRING, J. C., C. J. McCARTHY & A. J. WHITTEN. "1989" 1990. The terrestrial herpetofauna of the Mentawei Islands, Indonesia. *Indo-Malayan Zool.* 6: 119–132.

DUBOIS, A. 1980. Notes sur la systematique et la repartition des amphibiens anoures de Chine et des regions avoisinantes. IV. Classification generique et subgenerique de Pelobatidae, Megophryinae. *Bull. Mens. Soc. Linn., Lyon* 49(8): 469–482.

DUBOIS, A. 1982. *Leptophryne* Fitzinger, 1843, a senior synonym of *Cacophryne* Davis, 1935 (Bufonidae). *J. Herpetol.* 16: 173–174.

DUBOIS, A. 1982. Le statut nomenclatural des noms génériques d'amphibiens anoures créés par Kuhl & van Hasselt (1822): *Megophrys, Occidozyga* et *Rhacophorus. Bull. Mus. Nat. His. nat. Paris ser. 4* (sec A; nos. 1–2): 261–280.

DUBOIS, A. 1983. Classification et nomenclature supragénérique des Amphibiens Anoures. *Bull. Soc. Linn. Lyon* 52: 270–276.

DUBOIS, A. 1986. Miscellanea taxomonica batrachologica (I). *Alytes* 5(1–2): 7–95.

DUBOIS, A. 1987 "1986". Living amphibians of the world: a first step towards a comprehensive checklist. *Alytes* 5(3): 99–149.

DUBOIS, A. 1987. Again on the nomenclature of frogs. *Alytes* 6(1–2): 27–55.

DUBOIS, A. 1988. *Hyla reinwardtii* Schlegel, 1840(?) (Amphibia, Anura): proposed conservation. *Alytes* 7(3): 101–104.

DUBOIS, A. 1992. Notes sur la classification des Ranidae (Amphibiens Anoures). *Alytes* 61(10): 305–352.

DUBOIS, A. 1992. *Megophrys montana* Kuhl & van Hasselt, 1822 (Amphibia, Anura): proposed placement of both the generic and specific names on Official Lists, and *Leptobrachium parvum* Boulenger, 1893 (currently *Megophrys parva*): proposed conservation of the specific name. *Bull. Zool. Nom.* 49(3): 213–216.

DUBOIS, E. 1892. Voorloopig bericht omtrent het onderzoek naar de Pleistocene en Tertiaire vertebraten-fauna van Sumatra en Java, gedurende het jaar 1890. *Natuur. Tijd. Ned-Indië* 12: 93–100.

DUBOIS, E. 1908. Das geologische Alter der Kendeng-oder Trinil Fauna. *Tijdschr. Kon. Ned. Aardr. Gen. Ser. 2* 25: 1235–1270.

DUELLMAN, W. E. 1979. The number of amphibians and reptiles. *Herpetol. Rev.* 10: 83–84.

DUELLMAN, W. E. 1993. Amphibian species of the world: additions and corrections. *Sp. Publ. Mus. nat. Hist. Univ. Kansas* 21: i–iii + 1–372.

DUELLMAN, W. E. & L. TRUEB. "1986" 1985. Biology of amphibians. McGraw-Hill, New York. xix + 670 pp.

DUMÉRIL, A.-M.-C. 1853. Prodrome de la classification des reptiles ophidiens. *Mem. Acad. Sci.* 23: 399–536; 2 pl.

DUMÉRIL, A.-M.-C & G. BIBRON. 1836. Erpétologie Générale ou Histoire Naturelle compléte des Reptiles. Vol. 3. Librarie Encyclopedique de Roret, Paris. iv + 517 pp + errata et emendanda.

DUMÉRIL, A.-M.-C & G. BIBRON. 1837. Erpétologie Générale ou Histoire Naturelle compléte des Reptiles. Vol. 4. Librarie Encyclopedique de Roret, Paris. ii + 571 pp + errata et emendanda.

DUMÉRIL, A.-M.-C., G. BIBRON & A.-H.-A. DUMÉRIL. 1854a. Erpétologie Générale ou Histoire Naturelle compléte des Reptiles. Vol. 7. Part 1. Librarie Encyclopedique de Roret, Paris. xvi + 780 pp.

DUMÉRIL, A.-M.-C., G. BIBRON & A.-H.-A. DUMÉRIL. 1854b. Erpétologie Générale ou Histoire Naturelle compléte des Reptiles. Vol. 7. Part 2. Librarie Encyclopedique de Roret, Paris. xii + 781–1536 (= 756 pp).

DUMÉRIL, A.-M.-C., G. BIBRON & A.-H.-A. DUMÉRIL. 1854c. Erpétologie Générale ou Histoire Naturelle compléte des Reptiles. Atlas. Librarie Encyclopedique de Roret, Paris. 24 pp + Pl. 1–108.

DUNN, E. R. 1923. On a collection of reptiles from Sarawak. *J. Malay Br. Roy. Asiatic Soc.* 17(1): 1–4.

DUNN, E. R. 1927. Results of the Douglas Burden Expedition to the Island of Komodo. I.-Notes on *Varanus komodoensis*. *American Mus. Nov.* (286): 1–10.

DUNN, E. R. 1927. Results of the Douglas Burden Expedition to the Island of Komodo. II.-Snakes from the East Indies. *American Mus. Nov.* (287): 1–7.

DUNN, E. R. 1927. Results of the Douglas Burden Expedition to the Island of Komodo. III.-Lizards from the East Indies. *American Mus. Nov.* (288): 1–4.

DUNN, E. R. 1928. Results of the Douglas Burden Expedition to the Island of Komodo. IV.-Frogs from the East Indies. *American Mus. Nov.* (315): 1–9.

DUNSON, W. A. 1975. Adaptations of sea snakes. *In*: The biology of sea snakes. pp: 3–19. W. A. Dunson (Ed.). University Park Press, Baltimore.

EDELING, A. C. J. 1864. Reptiliën van Batavia en Borneo's westkust. *Natuur. Tijd. Ned-Indië* 26(1): 433.

EDELING, A. C. J. 1864. Slangen van Martapoera. *Natuur. Tijd. Ned-Indië* 26(1): 446.

EDELING, A. C. J. 1864. Slangen van Djokjakarta. *Natuur. Tijd. Ned-Indië* 26(1): 457.

EDELING, A. C. J. 1864. Recherches sur la faune erpétologique de Bornéo. *Natuur. Tijd. Ned-Indië* 26(1): 482.

EDELING, A. C. J. 1864. Reptiliën van Lahat. *Natuur. Tijd. Ned-Indië* 27(2): 387–388.

EDELING, A. C. J. 1864. Reptiliën in de Lampongsche distrikten verzameld door R. W. Diebel. *Natuur. Tijd. Ned-Indië* 26(1): 425.

EDELING, A. C. J. 1870. Recherches sur la faune erpétologique de Sumatra. *Natuur. Tijd. Ned-Indië* 31(1): 376–388.

EDGAR, P. W. & R. P. H. LILLEY. 1993. Herpetofauna survey of Manusela National Park. *In:* Natural history of Seram. pp: 131–141. I. D. Edwards, A. A. MacDonald & J. Proctor (Eds.). Intercept, Ltd., Andover.

EDWARDS, J. & J. DEAS. 1993. Notes on reptile exploitation in Java. *The Herpetile* 14(4): 77–88.

EDWARDS, J. & J. DEAS. 1996. The monitors seen during a short trip to Singapore and West Java. *Varanews* 4(4): 3–4.

ELKIN, J. 1991. Mass movement of turtles off the coast of Borneo. *Mar. Turtle Newsl.* (53): 16.

ELLEN, R. F., A. F. STIMSON & J. MENZIES. 1976. Structure and inconsistency in Nuaulu categories for amphibians. *J. Agric. Trop. Bot. Appl.* 23: 125–138.

ELLEN, R. F., A. F. STIMSON & J. MENZIES. 1977. The content of categories and experience; the case for some Nuaulu reptiles. *J. Agric. Trop. Bot. Appl.* 24: 3–22.

EMERSON, S. B. 1991. The ecomorphology of Bornean tree frogs (Family Rhacophoridae). *Zool. J. Linn. Soc.* 101: 337–357.

EMERSON, S. B. 1992. Courtship and nest-building behavior of a Bornean frog, *Rana blythi. Copeia* 1992(4): 1123–1127.

EMERSON, S. B. 1993. The comparative ecology of voiced and voiceless Bornean frogs. *J. Herpetol.* 26: 482–489.

EMERSON, S. B. 1996. Phylogenies and physiological processes—the evolution of sexual dimorphism in southeast Asian frogs. *Syst. Biol.* 45(3): 278–289.

EMERSON, S. B. & D. BERRIGAN. 1993. Systematics of southeast Asian ranids: multiple origins of voicelessness in the subgenus *Limnonectes* (Fitzinger). *Herpetologica* 49(1): 22–31.

EMERSON, S. B., Y. M. EAN & Y. K. LEE. 1990. On the jumping ability of leaf-mimicking frogs. *Sabah Mus. & Arch. J.* 1(3): 35–40.

EMERSON, S. B. & R. F. INGER. 1992. The comparative ecology of voiced and voiceless Bornean frogs. *J. Herpetol.* 26(4): 482–490.

EMERSON, S. B. & M. A. R. KOEHL. 1990. The interaction of behavioral and morphological change in the evolution of a novel locomotor type: "flying" frog. *Evolution* 44(8): 1931–1946.

EMERSON, S. B., C. N. ROWSEMITT & D. L. HESS. 1993. Androgen levels in a Bornean voiceless frog, *Rana blythi. Canadian J. Zool.* 71: 196–203.

EMERSON, S. B. & H. K. VORIS. 1992. Competing explanation for sexual dimorphism in a voiceless Bornean frog. *Functional Ecol.* 6: 654–660.

ENGKAMAT, L. A. 1987. Kajian ekologi ular laut (*Laticauda colubrina*) di Pulau Kalampunian Damit, Sabah. B. Sc. dissertation, Universiti Kebangsaan, Kota Kinabalu.

ENGKAMAT, L. A., R. B. STUEBING & H. K. VORIS. 1991. A population size estimate of the yellow-lipped sea krait, *Laticauda colubrina*, on Kalampunian Damit Island, Sabah, Malaysia. *Copeia* 1991(4): 1139–1142.

ERDELEN, W. 1988. Survey of the status of the water monitor (*Varanus salvator*, Reptilia: Varanidae) in south Sumatra. Report to the CITES Secretariat. 32 pg.

ERDELEN, W. 1991. Conservation and population ecology of monitor lizards: the water monitor *Varanus salvator* (Laurenti, 1768) in south Sumatra. *Mertensiella* (2): 120–135.

ERFTEMEIJER, P. & (P.) BOEADI. 1991. The diet of *Microhyla heymonsi* Vogt (Microhylidae) and *Rana chalconata* Schlegel (Ranidae) in a pond on west Java. *Raffles Bull. Zool.* 39(2): 279–282.

ERNST, C. H., R. W. BARBOUR & J. E. LOVICH. 1994. Turtles of the United States and Canada. Smithsonian Institution Press, Washington, D.C. xxxviii + 578 pp.

ESCHSCHOLTZ, J. 1829. Zoologischer Atlas. enthaltend Abbildungen und Beschreibungen neuer Thierarten, während des Flottcapitains von Kotzebue zweiter. Reise um die Welt, auf der Russisch-Kaiserlichen Kriegsschlupp Prepriaetië in den Jahren 1823–1826. Part I. Berlin. Reprinted 1966, Facsimile Reprint Series, The Ohio Herpetological Society, Athens, Ohio.

EVEN, E. 1989. Care and repeated breeding of the green ratsnake (*Gonyosoma oxycephala*). *Litteratura Serpentium* 9(4): 145–155.

EYDOUX, J. F. & F.-L.-P. GERVAIS. 1837. Reptiles. *Magasin Zool. J.* 7: 1–10.

FEJAVARY, S. J. 1927. Agro kozlemenyeh mit tudunk Z. Komodo ories gyi Krob. *Lermosgettindomanyi Kozlony* 59(3): 1–176, 283–285.

FISCHER, J. G. 1885. Ueber eine kollektion von Amphibien und Reptilien aus Sudost-Borneo. *Ark. Naturgesch., Berlin* 51: 41–72.

FITZINGER, L. 1826. Neue Classification der Reptilien nach ihren Natürlichen Verwandtschaften. Nebst einer Verwandtschafts-Tafel und einem Verzeichnisse der Reptilien-Sammlung des k. k. zoologischen Museums zu Wien. J. G. Huebner, Wien (= Vienna). viii + 66 pp.

FITZINGER, L. 1843. Systema reptilium. Fasciculus primus. Amblyglossae. Braunmuller & Seidel Bibliopolas, Vindobonae. 106 + ix pp.

FOEKEMA, G. G. M. 1977. Waargenomen amfibieën en reptielen tij dens een vakantie in Singapore en Indonesië. *Lacerta* 35: 35–43.

FOEKEMA, G. G. M. & P. VERHAART. 1977. Op zoek naar slangen op Java, Madura en Sulawesi. *Lacerta* 35: 151–172.

FORCART, L. 1953. Die Amphibien und Reptilien von Sumba, ihre zoogeographischen Beziehungen und Revision der Unterarten von *Typhlops polygrammicus*. *Verh. Naturf. Ges. Basel* 64(2): 356–388.

FRETEY, J. & R. BOUR. 1980. Redécouverte du type de *Dermochelys coriacea* (Vandelli)(Testudinata, Dermochelyidae). *Boll. Zool.* 47: 193–205.

FROGNER, K. J. & R. F. INGER. 1979. New species of narrow-mouth frogs (genus *Microhyla*) from Borneo. *Sarawak Mus. J.* 27: 311–322.

FROST, D. R. (Ed.) 1985. Amphibian species of the world. A taxonomic and geographical reference. Allen Press, Inc., and Assoc. Syst. Collections, Lawrence. v + 732 pp.

FUHN, I. 1969. Revision and redefinition of the genus *Ablepharus* Lichtenstein, 1823 (Reptilia, Scincidae). *Rev. Roum. Biol. Ser. Zool.* 14: 23–41.

FUHN, I. 1969. The polyphletic origin of the genus *Ablepharus* (Reptilia, Scincidae): a case of parallel evolution. *Z. Zool. Syst. Evolutionsforsch.* 7: 67–76.

FUKADA, H. 1964. A small collection of snakes of the Kyoto University Borneo Expedition, 1963–1964. *Bull. Kyoto Gakugei Univ. B* 25: 71–77.

FUNK, R. S. & P. R. VILARO. 1980. An English translation of Robert Mertens' keys to the monitor lizards, with a list of currently recognised species and subspecies. *Bull. Chicago Herpetol. Soc.* 15(2): 31–46.

GALSTAUN, B. 1973. Eiablagen des Komodowarens (*Varanus komodoensis*) im Zoologischen und Botanischen Garten, Jakarta. *Zool. Gart.* 43: 136–139.

GANS, C. 1955. Localities of the herpetological collections made during the "Novara Reise". *Ann. Carnegie Mus.* 33: 275–285.

GATESY, J. 1992. Sequence similarity of 12S ribosomal segment of mitochondrial DNAs of gharial and false gharial. *Copeia* 1992(1): 241–243.

GAULKE, M. 1996. On the herpetofauna of Palawan (Philippines). *Reptilian* 4(3): 26–32.

GAULKE, M. & U. FRITZ. 1997. Distribution patterns of batagurid turtles in the Philippines. *In*: Herpetology '97. Abstracts of the Third World Congress of Herpetology. 2–10 August, 1997. pp: 74–75. Z. Rocek & S. Hart (Eds). Third World Congress of Herpetology, Prague.

GAULKE, M. & U. FRITZ. 1997. On the turtle fauna of northern Sumatra (Sumatera Utara). *In*: Herpetology '97. Abstracts of the Third World Congress of Herpetology. 2–10 August, 1997. pp: 75. Z. Rocek & S. Hart (Eds). Third World Congress of Herpetology, Prague.

GEBAUR, S. 1993. Komodo dragon. *Muhibah* March/April, 1993: 43–44.

GEOFFROY-ST. HILLAIRE, E. F. 1809. Sur les tortues molles, nouveau genre sous le nom de *Trionyx* et sur la formation des carapaces. *Ann. Mus. Hist. Nat. Paris* 14: 1–20.

GERICKE, F. 1984. *Gonocephalus grandis*—Erfolgreiche Haltung und Nachzucht. *Sauria, Berlin* 6(4): 5–6.

GIBSON-HILL, C. A. 1950. A note on the reptiles occurring on the Cocos-Keeling Islands. *Bull. Raffles Mus.* 22: 206–211.

GIESEN, W. 1993. Mass feeding by dog-faced water snakes (*Cerberus rhynchops*) on Sumatran mudflats. *Malayan nat. J.* 46(3–4): 265–266.

GIRARD, C. 1857. Descriptions of some new reptiles, collected by the U.S. Exploring Expedition, under the command of Capt. Charles Wilkes, U.S. N. Third Part.—Including the species of Ophidians, exotic to North America. *Proc. Acad. Nat. Sci. Philadelphia* 9: 181–182.

GISTEL, J. 1848. Naturgeschichte des Thierreichs. Für hohere Schulen. Hoffman, Stuttgart. xvi + 216 pp + 32 pl.

GMELIN, J. F. 1789. Systema naturae. Editio decima tertia, aucta reformata. Stockholm. 1(3): 1031–1516.

GOLAY, P. 1985. Checklist and keys to the terrestrial proteroglyphs of the world (Serpentes: Elapidae—Hydrophiidae). Elapsoidea, Genèva. ix + 91 pp.

GOLAY, P., H. M. SMITH, D. G. BROADLEY, J. R. DIXON, C. McCARTHY, J.-C. RAGE, B. SCHÄTTI & M. TORIBA. 1993. Endoglyphs and other major venomous snakes of the world, a checklist. Azemiops S. A., Herpetological Data Centre, Aïre-Geneva. xv + 478 pp.

GOLDER, F. 1987. Zur Haltung und Fortpflanzung von *Boiga cyanea* (Duméril & Bibron, 1854) und Angaben zum Farbdimorphismus zweischen juvenilen und adulten Tieren. *Salamandra* 23(2/3): 78–83.

GOPALAKRISHNAKONE, P. 1985. Structure of the spinous scales of *Lapemis hardwickii* (1) Light, transmission electron and scanning electron microscopic study. *The Snake* 17(2): 148–155.

GOPALAKRISHNAKONE, P. 1992. Structure of the sublingual salivary gland of the reticulated python, *Python reticulatus. The Snake* 23(1): 19–24.

GOPALAKRISHNAKONE, P. & L. M. CHOU. (Eds.). 1990. Snakes of medical importance. National University of Singapore, Singapore. v + 670 + vi pp.

GOPALAKRISHNAKONE, P. & E. KOCHVA. 1993. Histological features of the venom apparatus of sea snake *Lapemis curtus. The Snake* 25(1): 27–37.

GORHAM, S. W. 1966. Liste der rezenten Amphibien und Reptilien. Ascaphidae, Leiopelmatidea (sic), Pipidae, Discoglossidae, Pelobatidae, Leptodactylidae, Rhinophrynidae. Das Tierreich 85. Walter de Gruyter & Co. i–xvi + 222 pp.

GORHAM, S. W. 1974. Checklist of world's amphibians up to January 1, 1970. New Brunswick Museum, Saint John. 173 pp.

GRAVENHORST, J. L. C. 1807. Vergleichende Ueberlicht des Linneischen und einiger neuern Zoologischen Systems. Ben Heinrich Dietrich. Göttingen. xvi + 476 pp.

GRAVENHORST, J. L. C. 1829. Deliciae Musei Zoologici Vratislaviensis. Fasciculcus Primus continens Chelonios et Batrachia. Leopold Voss, Lipsiae (Leipzig). xiv + 106 pp; 17 pl.

GRAY, C., L. MARCUS, W. C. McCARTEN & T. SAPPINGTON. 1966. Amoebiasis in the Komodo dragon. *Int. Zoo Yrbk.* 6: 279–283.

GRAY, J. E. 1825. A synopsis of the genera of reptiles and Amphibia, with a description of some new species. *Ann. Philos. Ser. 2* 10: 193–217.

GRAY, J. E. 1831. Synopsis Reptilum; or short descriptions of the species of reptiles. Part I.—Cataphracta. Tortoises, crocodiles, and Enaliosaurians. Treuttel, Wurtz, and Co., London. viii + 85 pp.

GRAY, J. E. 1831. Description of two new genera of frogs discovered by John Reeves, Esq., in China. *Zool. Misc., London* 1: 38.

GRAY, J. E. 1831. A synopsis of the species of the Class Reptilia. *In*: The Class Reptilia, arranged by the Baron Cuvier, with specific descriptions. Vol. 9. pp: 1–110. E. Griffith & E. Pidgeon (Eds.). Whittaker, Treacher, and Co., London.

GRAY, J. E. 1834. Characters of several new species of freshwater tortoises (*Emys*) from India and China. *Proc. Zool. Soc. London* 1834: 53–54.

GRAY, J. E. 1834. Characters of two new genera of reptiles (*Geoemyda* and *Gehyra*). *Proc. Zool. Soc. London* 1834: 99–100.

GRAY, J. E. "1838" 1839. Catalogue of the slender-tongued saurians, with descriptions of many new genera and species. *Ann. & Mag. nat. Hist.* 2: 287–293.

GRAY, J. E. 1839. Catalogue of the slender-tongued saurians, with descriptions of many new genera and species. *Ann. & Mag. nat. Hist.* 2: 331–337.

GRAY, J. E. 1842. Descriptions of some new species of reptiles, chiefly from the British Museum collection. *Zool. Misc.* 1842: 57–59.

GRAY, J. E. 1844. Catalogue of the tortoises, crocodiles, and amphiesbænians in the collection of the British Museum. British Museum Natural History, London. 80 pp.

GRAY, J. E. 1845. Catalogue of the specimens of lizards in the collection of the British Museum. British Museum (Natural History), London. xxviii + 289 pp.

GRAY, J. E. 1846. Descriptions of some new Australian reptiles. *In*: Discoveries in Australia: with an account of the coasts and rivers explored and surveyed during the voyage of the H. M. S. Beagle, in the years 1837–38–39–40–41–42–43. by Command of the Lords Commissioners of the Admiralty and a narrative of Captain Owen Stanley's visits to the islands in the Arafura Sea. Vol. I. pp: 498–504; Pl. 3–4. J. Lort Stokes (author). T. and W. Boone, London.

GRAY, J. E. 1849. Catalogue of the specimens of snakes in the collection of the British Museum. British Museum, London. xv + 125 pp.

GRAY, J. E. 1849. Description of three new genera and species of snakes. *Ann. & Mag. nat. Hist. Ser. 2* 4(22): 246–248.

GRAY, J. E. 1854/1852. Description of a new genus and some new species of tortoises. *Proc. Zool. Soc. London* 1854/1852: 133–135.

GRAY, J. E. 1863. Observations on the box tortoises, with the description of three new Asiatic species. *Proc. Zool. Soc. London* 1863: 173–177.

GRAY, J. E. 1864. Notes on certain species of tortoises from the Asiatic islands transmitted to the British Museum by Dr. Bleeker. *Proc. Zool. Soc. London* 1864: 11–13.

GRAY, J. E. 1864. Revision of the species of Trionychidae found in Asia and Africa, with the descriptions of some new species. *Proc. Zool. Soc. London* 1864: 76–98.

GRAY, J. E. 1870. Supplement to the catalogue of shield reptiles in the collection of the British Museum. Part I. Testudinata (tortoises). British Museum, London. x + 120 pp.

GRAY, J. E. 1873. On a new freshwater tortoise from Borneo (*Orlitia borneensis*). *Ann. & Mag. nat. Hist. Ser. 4* 11: 156–157.

GRAY, J. E. & R. HARWICKE. 1830–1832. Illustrations of Indian zoology, chiefly selected from the collections of Major-General Hardwicke. Vol. 1 (Parts I–X). Treuttel, Wurtz, Treuttel Jr., & Richter, London. (1) + (2) + 12 plates. (according to Sawyer, 1953; and Zhao and Adler, 1993, the years of publication of the plates are as follows: Plates 72, 75, 77–78, 82–83: 1830; Plates 73–74, 76, 79–80: 1831; Plate 81: 1832.)

GREEN, B., D. KING, M. BRAYSHER & A. SAIM. 1991. Thermoregulations, water turnover and energetics of free-living Komodo Dragons, *Varanus komodoensis. Comp. Biochem. Physiol.* 99A(1/2): 97–101.

GREENE, H. W. 1989. Defensive behavior and feeding biology of the Asian mock viper, *Psammodynastes pulverulentus* (Colubridae), a specialized predator of scincid lizards. *Chinese Herpetol. Res.* 2: 21–32.

GREENE, H. W. 1989. Ecological, evolutionary, and conservation implications of feeding biology in Old World cat snakes, genus *Boiga* (Colubridae). *Proc. California Acad. Sci. Ser. 4* 46(8): 193–207.

GREENPEACE INTERNATIONAL. 1989. Sea turtles and Indonesia. 7th Conf. Parties to CITES, Lausanne. 9–20 October, 1989. 23 pg.

GREER, A. E. 1968. Clutch size in the scincid genus *Emoia. Copeia* 1968: 417–418.

GREER, A. E. 1970. A subfamilial classification of scincid lizards. *Bull. Mus. Comp. Zool.* 139: 151–183.

GREER, A. E. 1970. The relationships of the skinks referred to the genus *Dasia. Breviora* (348): 1–30.

GREER, A. E. 1974. The generic relationships of the scincid lizard genus *Leiolopisma* and its relatives. *Austr. J. Zool.* (Suppl. Ser.) 31: 1–67.

GREER, A. E. 1977. The systematics and evolutionary relationships of the scincid lizard genus *Lygosoma. J. Nat. Hist.* 11: 515–540.

GREER, A. E. 1985. The relationships of the lizard genera *Anelytropsis* and *Dibamus. J. Herpetol.* 19: 116–156.

GRIFFIN, L. E. 1909. A list of snakes found in Palawan. *Philippine J. Sci. Ser. D* 4(6): 595–601.

GRITIS, P. 1992. The red pipe snake. *Reptile & Amphib. Mag.* Jan./Feb., 1992: 14–17.

GRITIS, P. & H. K. VORIS. 1990. Variability and significance of parietal and ventral scales in the marine snakes of the genus *Lapemis* (Serpentes: Hydrophiidae), with comments on the occurrence of spiny scales in the genus. *Fieldiana Zool.* n.s. (56): i–iii + 1–13.

GROOMBRIDGE, B. & R. LUXMOORE. 1991. Pythons in south-east Asia. A review of distribution, status and trade in three selected species. CITES, Lausanne. 127 pp.

GROOMBRIDGE, B. & L. WRIGHT (compiled). 1982. The IUCN Amphibia-Reptilia Red Data Book. Part 1—Testudines Crocodylia Rhynchocephalia. IUCN, Gland. (4) + xiii + 426 pp.

GROSSMAN, A. B. 1955. Review of "The systematic position of *Lanthonotus* and the affinities of the anguinomorphan lizards". *Quart. Rev. Biol.* 30: 400.

GROSSMANN, W. 1986. *Callagur borneensis* (Schlegel & Müller). *Sauria, Berlin* 8(4): 1–2.

GROSSMANN, W. 1986. Erste Erfahrungen bei der Haltung und Nachzucht des olivfarbenen Baumskinkes *Dasia olivacea* Gray, 1838. *Sauria, Berlin* 8(4): 13–21.

GROSSMANN, W. 1987. *Cyrtodactylus rubidus* (Blyth). *Sauria, Berlin* 9(3): 2.

GROSSMANN, W. 1988. *Pedostibes hosii* (Boulenger). *Sauria, Berlin* 10(3): 1–2.

GROSSMANN, W. 1989. *Lycodon effraenis* Cantor. *Sauria, Berlin* 11(4): 1–2.

GROSMANN, W. 1990. *Bungarus flaviceps flaviceps* Reinhard. *Sauria, Berlin* 12(4): 1–2.

GROSSMANN, W. & W. MUDRACK. 1987. Beitrag zur Haltung und Nachzucht von *Gekko smithii* (Gray, 1842). *Sauria, Berlin* 9(3): 25–30.

GROSSMANN, W. & G. VOGEL. 1993. *Chrysopelea pelias. Sauria, Berlin* 15(1): 1–2.

GUIBÉ, J. 1953. Les batrachiens et les reptiles des régions Indo-Malaise et Australo-Neoguineenne. *C. R. Soc. Biogéogr.* 30: 257–268.

GUIBÉ, J. & R. ROUX-ESTÈVE. 1972. Les types de Schlegel (Ophidiens) présents dans les collections du Muséum d'Histoire Naturelle de Paris. *Zool. Med.* 47(9): 129–134.

GUMPRECHT, A. 1996. *Elaphe flavolineata* (Schlegel). *Sauria, Berlin* 18(3): 373–376.

GÜNTHER, A. C. L. G. 1858. Catalogue of colubrine snakes in the collection of the British Museum. British Museum, London. xvi + 281 pp.

GÜNTHER, A. C. L. G. 1859. Catalogue of the Batrachia Salientia in the collection of the British Museum. Taylor & Francis, London. xvi + 160 pp + Pl. I–XII.

GÜNTHER, A. C. L. G. 1862. On new species of snakes in the collection of the British Museum. *Ann. & Mag. nat. Hist. Ser. 3* 9: 124–132; Pl. IX–X.

GÜNTHER, A. C. L. G. 1863. Third account of new species of snakes in the collection of the British Museum. *Ann. & Mag. nat. Hist. Ser. 3* 12: 348–365; Pl. V–VI.

GÜNTHER, A. C. L. G. 1865. Fourth account of new species of snakes in the collection of the British Museum. *Ann. & Mag. nat. Hist. Ser. 3* 15: 89–98.

GÜNTHER, A. C. L. G. 1866. Sixth account of new species of snakes in the collection of the British Museum. *Ann. & Mag. nat. Hist. Ser. 3* 18: 24–29; Pl. VI–VII.

GÜNTHER, A. C. L. G. 1867. Additions to the knowledge of Australian reptiles and fishes. *Ann. & Mag. nat. Hist. Ser. 3* 20:45–68.

GÜNTHER, A. C. L. G. 1872. On the reptiles and amphibians of Borneo. *Proc. Zool. Soc., London* 1872: 586–600; Pl. XXXV–XXXX.

GÜNTHER, A. C. L. G. 1872. Seventh account of new species of snakes in the collection of the British Museum. *Ann. & Mag. nat. Hist. Ser. 4* 9: 13–37; Pl. III–VI.

GÜNTHER, A. C. L. G. 1873. Notes on some reptiles and batrachians obtained by Dr. Adolf Bernhard Meyer in Celebes and the Philippine Islands. *Proc. Zool. Soc. London* 1873: 165–172.

GÜNTHER, A. C. L. G. 1895. The reptiles and batrachians of the Natuna Islands. *Novit. Zool.* 2: 499–502.

GÜNTHER, R. & U. MANTHEY. 1995. *Xenophidion*, a new genus with two new species of snakes from Malaysia (Serpentes, Colubridae). *Amphibia-Reptilia* 16: 229–240.

GÜNTHER, R. & V. WALLACH. 1997. The systematic position of the Malaysian snake genus *Xenophidion. In:* Herpetology '97. Abstracts of the Third World Congress of Herpetology. 2–10 August, 1997. pp: 88. Z. Rocek & S. Hart (Eds). Third World Congress of Herpetology, Prague.

GYI, K. K. 1970. A revision of colubrid snakes of the subfamily Homalopsinae. *Univ. Kansas Publ. Mus. nat. Hist.* 20: 47–223.

HADDON, A. C. 1962. The tortoise and the mouse deer (Kenyah). *Sarawak Mus. J. n.s.* 10(19–20): 535–536.

HAHN, D. E. 1980. Liste der rezenten Amphibien und Reptilien. Anomalepidae, Leptotyphlopidae, Typhlopidae. Das Tierreich 101. Walter de Gruyter & Co. 1–93 pp.

HAILE, N. S. 1958. The snakes of Borneo, with a key to the species. *Sarawak Mus. J.* 8: 743–771.

HAILE, N. S. 1963. Snake bites man: Two recent Borneo cases. *Sarawak Mus. J.* 11: 291–298.

HAILE, N. S. & R. F. INGER. 1959. Two new frogs from Sarawak. *Sarawak Mus. J.* 9: 13–14.

HALLOWELL, E. 1861. Report upon the Reptilia of the North Pacific Exploring Expedition, under command of Capt. John Rogers U.S. N. *Proc. Acad. nat. Sci. Philadelphia* 1860(12): 480–509.

HAN, K. H., R. B. STUEBING & H. K. VORIS. 1991. Population structure and reproduction in the marine snake, *Lapemis hardwickii* Gray, from the west coast of Sabah. *Sarawak Mus. J.* n.s. 42(63): 463–475.

HAN, S. S. 1988. The reproductive ecology and diet of a sea snake (*Lapemis hardwickii* Gray, 1835) from west coast of Sabah. Unpublished M. Sc. dissertation, University Kebangsaan Malaysia, Sabah. 102 pg.

HANITSCH, R. 1900. An expedition to Mount Kina Balu, British North Borneo. *J. Str. Br. Royal Asiatic Soc.* (34): 48–88; 2 pl.

HARDING, K. A. 1982. Courtship display in a Bornean frog. *Proc. Biol. Soc. Washington* 95: 621–624.

HARDING, K. A. & K. R. G. WELCH. 1980. Venomous snakes of the world. A checklist. Pergamon Press, Oxford. x + 188 pp.

HARRISSON, B. 1961. *Lanthonotus borneensis*—habits and observations. *Sarawak Mus. J.* 10: 17–18.

HARRISSON, B. 1962. Beobachtungen am lebenden Taubwaran *Lanthonotus borneensis. Natur und Mus.* 92(2): 38–45. [English translation: Observations on living earless monitors (*Lanthonotus borneensis*) by P. Gritis. *Bull. Chicago Herpetol. Soc.* 24(10): 185–188.]

HARRISSON, J. L. 1956. The bite of a blue Malaysian coral snake or "ular matahari" *(Maticora bivirgata). Malay. nat. J.* 11: 130–132.

HARRISSON, J. L. 1960. Evolution of flying membranes. *Malay. nat. J.* 14: 132–134.

HARRISSON, T. 1950. Chick versus king cobra. *Sarawak Mus. J.* 5: 326–327.

HARRISSON, T. 1950. The Sarawak turtle islands' "Semah". *Malayan Br. Royal Asiatic Soc.* 23(3): 105–126.

HARRISSON, T. 1951. The edible turtle (*Chelonia mydas*) in Borneo: 1. Breeeding season. *Sarawak Mus. J.* 3: 593–596.

HARRISSON, T. 1954. The edible turtle (*Chelonia mydas*) in Borneo: 2. Copulation. *Sarawak Mus. J.* 6: 126–128.

HARRISSON, T. 1955. The edible turtle (*Chelonia mydas*) in Borneo: 3. Young turtles (in captivity). *Sarawak Mus. J.* 6: 633–640.

HARRISSON, T. 1956. The edible turtle (*Chelonia mydas*) in Borneo: 4. Growing turtles and growing problems. *Sarawak Mus. J.* 7: 233–239..

HARRISSON, T. 1956. The edible turtle (*Chelonia mydas*) in Borneo: 5. Tagging turtles (and why). *Sarawak Mus. J.* 7: 504–515.

HARRISSON, T. 1958. Notes on the edible turtle (*Chelonia mydas*): 6. Semah ceremonies, 1949–1958. *Sarawak Mus. J.* 8: 482–486.

HARRISSON, T. 1958. Notes on the edible turtle (*Chelonia mydas*): 7. Long-term tagging returns, 1952–1958. *Sarawak Mus. J.* 8: 772–774.

HARRISSON, T. 1959. Notes on the edible turtle (*Chelonia mydas*): 8. First tag returns outside Sarawak. *Sarawak Mus. J.* 9: 277–278.

HARRISSON, T. 1961. Niah's new cave gecko: habits. *Sarawak Mus. J.* 10: 277–282.

HARRISSON, T. 1961. Notes on the green turtle (*Chelonia mydas*): 9. Some new hatching observations. *Sarawak Mus. J.* 10: 293–299.

HARRISSON, T. 1962. Catsnake kills fruit bat. *Malay. nat. J.* 16(3): 153–154.

HARRISSON, T. 1962. Notes on the green turtle (*Chelonia mydas*): 10. Some emergence variations. 11. West Borneo numbers, the downward trend. 12: Monthly laying cycles. *Sarawak Mus. J.* 10: 610–630.

HARRISSON, T. 1963. *Lanthonotus borneensis*: the first 30 live ones. *Sarawak Mus. J.* 11: 299–301.

HARRISSON, T. 1963. Notes on marine turtles: 13. Growth rate of the hawksbill. 14: Albino green turtles and sacred ones. *Sarawak Mus. J.* 11: 302–306.

HARRISSON, T. 1964. Notes on marine turtles—15: Sabah's turtle islands. *Sarawak Mus. J.* 11: 624–627.

HARRISSON, T. 1965. Notes on marine turtles—16: Some loggerhead and hawksbill comparisons with the green turtle. *Sarawak Mus. J.* 12: 419–422.

HARRISSON, T. 1966. Cold-blooded vertebrates of the Niah Cave area. *Sarawak Mus. J.* 14: 276–286.

HARRISSON, T. 1966. Notes on marine turtles—17: Sabah and Sarawak islands compared. *Sarawak Mus. J.* 14: 335–340.

HARRISSON, T. 1967. Notes on marine turtles—18: A report on the Sarawak turtle industry (1966) with recommendations for the future. *Sarawak Mus. J.* 15: 424–436.

HARRISSON, T. 1969. The marine turtle situation in Sarawak. *IUCN Publ. (Occ. Pap.)* 20: 171–173.

HÄUPL, M., F. TIEDEMANN & H. GRILLITSCH. 1994. Kataloge der wissenschaftlichen Sammlungen des Naturhistorischen Museums in Wien. Vertebrata Heft 3. Katalog der Typen der Herpetologischen Sammlung nach dem Stand vom 1. Jänner 1994. Band 9. Teil I: Amphibia. Naturhistorisches Museum Wien, Vienna. 46 pp.

HECHT, M. K., C. KROPACH & B. M. HECHT. 1974. Distribution of the yellow-bellied sea snake, *Pelamis platyurus*, and its significance in relation to the fossil record. *Herpetologica* 30(4): 387–396.

HEDIGER, H. 1932. Zum Problem der "fliegenden" Schlangen. *Rev. Suisse Zool.* 39: 239–246.

HEIJNEN, G. H. 1989. The maintenance and breeding of the red-tailed green racer, *Gonyosoma oxycephala. Snake Keeper* 3(6): 9–11.

HEIJNEN, G. H. 1989. The maintenance and breeding of the red-tailed green racer, *Gonyosoma oxycephala.* Part 2. *Snake Keeper* 3(7): 4–7.

HENDERSON, R. W. & M. H. BINDER. 1980. The ecology and behavior of vine snakes (*Ahaetulla, Oxybelis, Thelotornis, Uromacer*): a review. *Milwaukee Pub. Mus. Contrib. Biol. Geol.* 37: 1–38.

HENDRICKSON, J. R. 1958. The green turtle, *Chelonia mydas* (Linn.) in Malaya and Sarawak. *Proc. Zool. Soc., London* 130: 455–566.

HENNE AM RHYNE, R. 1903. Einege mer Kuiürdige Kreichtiere der Sund-Inseln. *Zool. Anz.* 16: 167–172.

HENNIG, W. 1936. Revision der Gattung *Draco* (Agamidae). *Temminckia* 1: 153–220.

HERMANN, H.-J. 1986. Schweben und Gleiten bei Amphibien und Reptilien. *Sauria, Berlin* 8(1): 13–17.

HERMANN, H.-J. 1991. Herpetofocus. *Herpetofauna* 13: 11.

HERMANN, H.-J. & S. GERLACH. 1981. Der Bananenfrösch, *Rhacophorus leucomystax* (Gravenhorst, 1829). *Aquar. Terr.* 28: 432.

HERMANN, H.-W. & U. JOGER. "1997" 1996. Evolution of viperine snakes. *In*: Venomous snakes: ecology, evolution and snakebite. pp: 43–61. R. S. Thorpe, W. Wüster & A. Malhotra (Eds.). Symp. Zool. Soc. London (70). The Zoological Society of London/Clarendon Press, Oxford.

HERRE, A. W. 1958. On the gliding of flying lizards, genus *Draco. Copeia* 1958(4): 338–339.

HIKIDA, T. 1979. [Notes on the Bornean arboreal skink, *Apterygodon vittatus*.] *Japanese J. Herpetol.* 8: 68–69. [In Japanese.]

HIKIDA, T. 1980. [Lizards of Borneo.] *Acta Phytotaxonomica et Geobotanica* 31: (1–3): 97–102. [In Japanese.]

HIKIDA, T. 1982. A new limbless *Brachymeles* (Sauria: Scincidae) from Mt. Kinabalu, North Borneo. *Copeia* 1982(4): 840–844.

HIKIDA, T. & H. OTA. 1994. *Sphenomorphus aquaticus* Malkmus 1991, a junior synonym of *Tropidophorus beccarii* (Peters, 1871)(Reptilia: Squamata: Scincidae). *Bonner Zool. Beitr.* 45: 57–60.

HIN, H. K., R. B. STUEBING & H. K. VORIS. 1991. Population structure and reproduction in the marine snake, *Lapemis hardwickii* Gray from the west coast of Sabah. *Sarawak Mus. J.* 32: 463–475.

HIRAYAMA, R. 1984. Cladistic analysis of batagurine turtles (Batagurinae: Emydidae: Testudinoidea): A preliminary result. Studia Geologica Salmanticensia. Vol. Especial 1. *Studia Palaeocheloniologica* 1: 141–157.

HIRTH, H. F. 1971. Synopsis of biological data on the green turtle, *Chelonia mydas* (Linnaeus) 1758. *FAO Fish. Syn.* 85: 1–8, 19.

HIRTH, H. F. 1980a. *Chelonia. Cat. American Amphib.-Rept.* 248: 1–2.

HIRTH, H. F. 1980b. *Chelonia mydas. Cat. Amer. Amphib.-Rept.* 249: 1–4.

HISADA, H. 1966. Komodo dragon lays eggs. *Anim. in Zoos* 18: 259.

HODGES, R. 1993. Snakes of Java with special reference to East Java. *British Herpetol. Soc. Bull.* (43): 15–32.

HOFFMANN, P. 1994. Fund der seltenen *Rana palavanensis* am Mt. Kinabalu auf Borneo. *Salamandra* 30: 223–224.

HOFFMANN, P. 1995. Untersuchungen zur Anurenfauna des Mt. Kinabalu (Borneo). Teil I: Einführung, Familien Pelobatidae, Bufonidae, Microhylidae. *Sauria, Berlin* 17(1): 3–10.

HOFFMANN, P. 1995. Untersuchungen zur Anurenfauna des Mt. Kinabalu (Borneo). Teil II: Familien Ranidae, Rhacophoridae. *Sauria, Berlin* 17(3): 9–17.

HOFFMANN, P. 1995. Zur Kenntnis des Engmaulfrosches *Chaperina fusca* Mocquard, 1892 vom Mt. Kinabalu auf Borneo. *Herpetofauna* 17(96): 27–29.

HOGE, A. R. & S. A. R. W. DE LEMOS ROMANO. 1974. Notes on *Trimeresurus brongersmai* Hoge 1969 (Serpentes, Viperidae, Crotalinae). *Mem. Instituto Butantan* 38: 147–158.

HOGE, A. R. & S. A. R. W. DE LEMOS ROMANO-HOGE. 1981. Poisonous snakes of the world. Part I. Check list of the pit vipers Viperoidea, Viperidae, Crotalinae. *Mem. Instituto Butantan* 42–43: 179–310.

HOLZINGER TENEVER, H. 1916. *Calamaria borneensis* nov. subsp. *Zool. Anz.* 48: 33–34.

HOLZINGER TENEVER, H. 1917. Verzeichnis der von Shoede auf Ceylon und Sumatra gesammelten Reptilien. *Mitt. Zool. Mus. Berlin* 8: 425–454.

HOLZINGER TENEVER, H. 1919. Herpetologische Mitteilungen aus dem Museum für Naturkunde in Oldenburg, Gr. I. Collection Dr. Lamping, Sumatra. *Arch. Natur. (A)* 85(11): 91–89.

HOLZINGER TENEVER, H. 1920. Herpetologische Mitteilungen aus dem Zoologischen Museum in Berlin. Die von H. Mertens in Indonesien gesammelten Reptilien. *Arch. Natur. (A)* 85: 99–111.

HONEGGER, R. 1986. Zur pflege und langjahrigen Nachzucht von *Siebenrockiella crassicollis* (Gray, 1831). *Salamandra* 22(1): 1–10.

HOOGERWERF, A. 1948. Enkele waarnemingen bij jonge Komodo-Varanen (*Varanus komodoensis* Ouwens) in den gevangen staat van eistadium bij dize Reptilien. *Chron. Nat.* 104(2): 33–42.

HOOGERWERF, A. 1955. *Varanus komodoensis. Neth. Commissie voor Intl. Natuurbeschrming* (15): 47–54.

HOOGERWERF, A. 1972. *Varanus* (Reptilia, Sauria) from the Pleistocene of Timor. *Zool. Med.* 47(34): 445–448.

HOOGMOED, M. S. & C. R. CRUMLY. 1984. Land tortoise types in the Rijksmuseum van Natuurlijke Histoire with comments on nomenclature and systematics (Reptilia: Testudines: Testudinidae). *Zool. Med.* 58(1): 241–259.

HOOIJER, D. A. 1948. Pleistocene vertebrates from Celebes. II *Testudo margae* sp. nov. *Proc. Koninkl. Ned. Akad. Wet.* 51(9): 1169–1182.

HOOIJER, D. A. 1951. Pygmy elephant and giant tortoise. *Sci. Monthly* 72(1): 3–8.

HOOIJER, D. A. 1954. Pleistocene vertebrates from Celebes. X. Testudinata. *Proc. Koninkl. Ned. Akad. Wet. B* 57(4): 486–489.

HOOIJER, D. A. 1971. A giant land tortoise *Geochelone atlas* (Falconer and Cautley), from the Pleistocene of Timor. I. *J. Koninkl. Ned. Akad. Wet. B* 74(5): 504–525.

HOOIJER, D. A. 1972. *Varanus* (Reptilia, Sauria) from the Pleistocene of Timor. *Zool. Med.* 47: 445–448.

HOOIJER, D. A. 1982. The extinct giant land tortoise and the pygmy stegodont of Indonesia. *Mod. Quart. Res. Southeast Asia* 7: 171–176.

HORN, H.-G. 1985. Beiträg zum Verhalten von Waranen: Die Ritualkämpfe von *Varanus komodoensis* (Ouwens 1912). *Salamandra* 21(2/3): 169–179.

HORN, H.-G. 1994. New data on ritualized combats in monitor lizards (Sauria: Varanidae), with remarks on their function and phylogenetic implications. *Zool. Garten N.F.* 64(5): 265–280.

HORN, H.-G. & B. SCHULZ. 1977. [*Varanus dumerili*, as few know it.] *Das Aquar.* 91: 37–38.

HORN, H.-G. & G. J. VISSER. 1991. Basic data on the biology of monitors. *Mertensiella* 2: 176–187.

HORNSTEDT, C. F. 1787. Beschreifning på en Ny Orm fråm Java. *Kon. Vet. Akad. Handl. Stockholm Ser. 2* 8: 306–308.

HORST, O. 1926. *Varanus komodoensis. Trop. Natuur* 15: 118–121.

HORTON, D. R. 1973. A new scincid genus from southeast Asia. *J. Herpetol.* 7(3): 283–287.

HORTON, D. R. 1973. Evolution of the genus *Mabuya* (Lacertilia, Scincidae). Unpublished Ph.D. thesis, University of New England, Armidale. 13 unnumbered + 311 pg.

HOW, R. A., L. H. SCHMITT & (?) MAHARADATUNKAMSI. 1996. Geographical variation in the genus *Dendrelaphis* (Serpentes: Colubridae) within the islands of south-east Indonesia. *J. Zool. London* 238: 351–363.

HUBRECHT, A. A. W. 1879. Contributions to the herpetology of Sumatra. *Notes Leyden Mus.* 1(4): 243–245.

HUBRECHT, A. A. W. 1881. On a new genus and species of Agamidae from Sumatra. *Notes Leyden Mus.* 3: 51–52.

HUBRECHT, A. A. W. 1887. Kruipende Dieren en Visschen-Systematische lijst. 1. Reptilia. *Natuurlijke Historie, Tweede Afdeeling* 4(1): 1–8.

HUSSAIN, S. A. 1984. Some aspects of the biology and ecology of Narcondam hornbill (*Rhyticeros narcondami*). *J. Bombay nat. Hist. Soc.* 81(1): 1–18.

HUTABARAT, H. P., I. S. SUWELO & S. SILALAHI. 1994. Hawksbill turtle in Kepulauan Seribu Marine National Park, Jakarta Bay. *Mar. Turtle Newsl.* (64): 12–13.

IN DEN BOSCH, H. A. J. 1985. Snakes of Sulawesi: checklist, key and additional biogeographical remarks. *Zool. Verh.* 217: 1–50.

IN DEN BOSCH, H. A. J. & I. INEICH. 1994. The Typhlopidae of Sulawesi (Indonesia): a review with description of a new genus and a new species (Serpentes: Typhlopidae). *J. Herpetol.* 28(2): 206–217.

INEICH, I. 1995. Das Tierreich. The animal kingdom. Part 109. Familia Gekkonidae (Reptilia, Sauria) I. Australia and Oceania [Book Review]. *Herpetol. Rev.* 26(3): 159–162.

INEICH, I. & J. DEUVE. 1990. *Psammophis condanarus* (Asian sand snake). *Herpetol. Rev.* 21(1): 23.

INGER, R. F. 1954. Systematics and zoogeography of Philippine Amphibia. *Fieldiana Zool.* 33: 183–531.

INGER, R. F. 1954. On a collection of amphibians from Mount Kina Balu, North Borneo. *J. Wash. Acad. Sci.* 44: 250–251.

INGER, R. F. 1956. Some amphibians from the lowlands of North Borneo. *Fieldiana Zool.* 34: 389–424.

INGER, R. F. 1957. A new gecko of the genus *Cyrtodactylus* with a key to the species from Borneo and the Philippines islands. *Sarawak Mus. J.* 8: 261–264.

INGER, R. F. 1958. A new toad from Sarawak. *Sarawak Mus. J.* 8: 476–478.

INGER, R. F. 1958. Notes on the Bornean glass snake. *Sarawak Mus. J.* 8: 479–481.

INGER, R. F. 1958. Three new species related to *Sphenomorphus variegatus* (Peters). *Fieldiana Zool.* 39: 257–268.

INGER, R. F. 1959. Temperature responses and ecological relations of two Bornean lizards. *Ecol.* 40(1): 127–136.

INGER, R. F. 1960. Notes on the toads of the genus *Pelophryne*. *Fieldiana Zool.* 39: 415–418.

INGER, R. F. 1960. A review of the Oriental toads of the genus *Ansonia*. *Fieldiana Zool.* 39: 473–503.

INGER, R. F. 1962. Rare lizard reaches Museum. *Mus. News, Chicago* March, 1962: 7.

INGER, R. F. 1966. The systematics and zoogeography of the Amphibia of Borneo. *Fieldiana Zool.* 52: 1–402. Reprinted 1990. Lun Hing Trading Company, Kota Kinabalu.

INGER, R. F. 1967. A new colubrid snake of the genus *Stegonotus* from Borneo. *Fieldiana Zool.* 51(5): 77–83.

INGER, R. F. 1969. Organization of communities of frogs along small rain forest streams in Sarawak. *J. Anim. Ecol.* 38: 123–148.

INGER, R. F. 1972. *Bufo* of Eurasia. *In*: Evolution in the genus *Bufo*. pp: 1020–118. W. F. Blair (Ed.). University of Texas Press, Austin and London.

INGER, R. F. 1978. Frogs and toads. *In:* Kinabalu: Summit of Borneo. M. Luping, C. Wen & E. R. Dingley (Eds.). pp: 311–320. The Sabah Society, Kota Kinabalu.

INGER, R. F. 1980. Densities of floor-dwelling frogs and lizards in lowland forests in southeast Asia and central America. *American Natural.* 115: 761–770.

INGER, R. F. 1983. Larvae of southeast Asian species of *Leptobrachium* and *Leptobrachella* (Anura: Pelobatidae). *In:* Advances in herpetology and evolutionary biology. pp: 13–22. A. G. J. Rhodin & K. Miyata (Eds.). Museum of Comparative Zoology, Cambridge, Mass.

INGER, R. F. 1983. Morphological and ecological variation in the flying lizards (genus *Draco*). *Fieldiana Zool. n.s.* (18): i–iv + 1–35.

INGER, R. F. 1985. Tadpoles of forested regions of Borneo. *Fieldiana Zool.* 26: 1–89.

INGER, R. F. 1986. Diets of tadpoles living in a Bornean rain forest. *Alytes* 5(4): 153–164.

INGER, R. F. 1989. Four new species of frogs from Borneo. *Malayan nat. J.* 42: 229–243.

INGER, R. F. 1992. Variation of apomorphic characters in stream-dwelling tadpoles of the bufonid genus *Ansonia* (Amphibia: Anura). *Zool. J. Linn. Soc.* 105: 225–237.

INGER, R. F. 1992. A bimodal feeding system in a stream-dwelling larva of *Rhacophorus* from Borneo. *Copeia* 1992(3): 887–890.

INGER, R. F. 1996. Commentary on a proposed classification of the family Ranidae. *Herpetologica* 52(2): 241–246.

INGER, R. F. & J. P. BACON. 1968. Annual reproduction and clutch size in rainforest frogs from Sarawak. *Copeia* 1968(3): 602–606.

INGER, R. F., (P.) BOEADI & A. TAUFIK. 1996. New species of ranid frogs (Amphibia: Anura) from Central Kalimantan, Borneo. *Raffles Bull. Zool.* 44(2): 363–369.

INGER, R. F. & W. C BROWN. 1980. Species of the scincid genus *Dasia* Gray. *Fieldiana Zool.* 3: 1–11.

INGER, R. F. & J. DRING. 1988. Taxonomic and ecological relations of Borneo stream toads allied to *Ansonia leptopus* (Günther)(Anura: Bufonidae). *Malayan nat. J.* 41(4): 461–471.

INGER, R. F., Y. M. EAN & Y. K. LEE. 1987. Key to the frogs of the Danum Valley Field Centre. *Sabah Soc. J.* 8(3): 373–379.

INGER, R. J. & K. J. FROGNER. 1980. New species of narrow-mouth frogs (genus *Microhyla*) from Borneo. *Sarawak Mus. J.* 27: 311–324.

INGER, R. F. & B. GREENBERG. 1963. The annual reproductive pattern of the frog *Rana erythraea* in Sarawak. *Physiol. Zool.* 36: 21–33.

INGER, R. F. & B. GREENBERG. 1966. Ecological and competitive relations among three species of frogs (genus *Rana*). *Ecol.* 47: 746–759.

INGER, R. F. & P. A. GRITIS. 1983. Variation in Bornean frogs of the *Amolops jerboa* species group, with description of two new species. *Fieldiana Zool.* 19: 1–13.

INGER, R. F. & A. E. LEVITON. 1961. A new colubrid snake of the genus *Pseudorabdion* from Sumatra. *Fieldiana Zool.* 44: 45–57.

INGER, R. F. & A. E. LEVITON. 1966. The taxonomic status of Bornean snakes of the genus *Pseudorabdion* Jan and of the nominal genus *Idiopholis* Mocquard. *Proc. California Acad. Sci.* 34(4): 307–314.

INGER, R. F. & H. MARX. 1965. The systematics and evolution of the Oriental colubrid snakes of the genus *Calamaria. Fieldiana Zool.* 49: 1–304.

INGER, R. F. & R. STUEBING 1989. Frogs of Sabah. Sabah Parks Trustees, Kota Kinabalu. 132 + iv pp.

INGER, R. F. & R. STUEBING. 1991. A new species of frog of the genus *Leptobrachella* Smith (Anura: Pelobatidae) with a key to the species from Borneo. *Raffles Bull. Zool.* 39(1): 99–103.

INGER, R. F. & R. STUEBING. 1992. The montane amphibian fauna of northwestern Borneo. *Malayan nat. J.* 46: 41–51.

INGER, R. F. & R. STUEBING. 1994. First record of the lizard genus *Pseudocalotes* (Lacertilia; Agamidae) in Borneo, with description of a new species. *Raffles Bull. Zool.* 42(4): 961–965.

INGER, R. F. & R. STUEBING. 1996. Two new species of frogs from southeastern Sarawak. *Raffles Bull. Zool.* 44(2): 543–549.

INGER, R. F., R. B. STUEBING & F. L. TAN. 1995. New species and new records of anurans from Borneo. *Raffles Bull. Zool.* 43(1): 115–131.

INGER, R. F. & F. L. TAN. 1990. Recently-discovered and newly-assigned frog larvae (Ranidae and Rhacophoridae) from Borneo. *Raffles Bull. Zool.* 38(1): 3–9.

INGER, R. F. & F. L. TAN. 1995. New species and new records of anurans from Borneo. *Raffles Bull. Zool.* 43(1): 115–131.

INGER, R. F. & F. L. TAN. 1996. Checklist of the frogs of Borneo. *Raffles Bull. Zool.* 44(2): 551–574.

INGER, R. F. & F. L. TAN. 1996. The natural history of amphibians and reptiles of Sabah. Natural History Publications (Borneo), Sdn. Bhd., Kota Kinabalu. vi + 101 pp.

INGER, R. F. & H. K. VORIS. 1988. Taxonomic status and reproductive biology of Bornean tadpole-carrying frogs. *Copeia* (4): 1060–1061.

INGER, R. F. & H. K. VORIS. 1992. Variation in communities of amphibians in Bornean forests. *In*: Forest biology and conservation in Borneo. pp: 494–495. G. Ismail, M. Mohamed & S. Omar (Eds.). Centre for Borneo Studies Publication No. 2. Yayasan Sabah, Kota Kinabalu.

INGER, R. F. & H. K. VORIS. 1993. A comparison of amphibian communities through time and from place to place in Bornean forests. *J. Trop. Ecol.* 9: 409–433.

INGER, R. F., H. K. VORIS & K. J. FROGNER. 1986. Organisation of a community of tadpoles in rainforest streams in Borneo. *J. Trop. Ecol.* 2: 193–205.

INGER, R., F., H. K. VORIS & H. H. VORIS. 1974. Genetic variation and population ecology of some southeast Asian frogs of the genera *Bufo* and *Rana*. *Biochem. Genetics* 12: 121–145.

INGER, R. F., H. K. VORIS & P. WALKER 1985. A key to the frogs of Sarawak. *Sarawak Mus. J.* 34: 161–182.

INGER, R. F., H. K. VORIS & P. WALKER 1986. Larval transport in a Bornean ranid frog. *Copeia* 1986(2): 523–525.

INGER, R. F. & R. J. WASSERSUG. 1990. A centrolenid-like anuran larvae from southeast Asia. *Zool. Sci.* 7: 557–561.

INTERNATIONAL COMMISSION OF ZOOLOGICAL NOMENCLATURE. 1986. Opinion 1374. *Boiga* Fitzinger, 1826 (Reptilia, Serpentes): conserved. *Bull. Zool. Nom.* 435(1): 25–26.

ISKANDER, D. T. 1978. A new species of *Barbourula*: First record of a discoglossid anuran from Borneo. *Copeia* 1978: 564–566.

ISKANDER, D. T. 1979. A second specimen of the Matanna water snake, *Enhydris matannensis* (Blgr), from Raha, Muna Island, Indonesia. *British J. Herpetol.* 5: 849–850.

ISKANDER, D. T. 1986. On the occurrence of *Trimeresurus monticola* in Sumatra. Confirmation of a century doubtful snake record. *The Snake* 18(2): 126–128.

ISKANDER, D. T. 1987. The occurrence of *Enhydris alternans* at Java. *The Snake* 19(1): 72–73.

ISKANDER, D. T. 1987. A new colour variation of *Gonyosoma oxycephalum* from central Java. *The Snake* 19(2): 129–132.

ISKANDER, D. T. 1995. Note on the second specimen of *Barbourula kalimantanensis* (Amphibia: Anura: Discoglossidae). *Raffles Bull. Zool.* 43(2): 309–311.

ISKANDER, D. T., (P.) BOEADI & M. SANCOYO. 1996. *Limnonectes kadarsani* (Amphibia: Anura: Ranidae), a new frog from the Nusa Tenggara Islands. *Raffles Bull. Zool.* 44(1): 21–28.

IVERSON, J. B. 1986. A checklist with distribution maps of the turtles of the world. Privately printed, Richmond, Indiana. viii + 283 pp.

IVERSON, J. B. 1992. A revised checklist with distribution maps of the turtles of the world. Privately printed, Richmond, Indiana. xiii + 363 pp.

JACKSON, K. & T. H. FRITTS. 1995. Evidence from tooth surface morphology for a posterior maxillary origin of the proteroglyph fang. *Amphibia-Reptilia* 16: 273–288.

JACKSON, K. & T. H. FRITTS. 1996. Observations of a grooved anterior fang in *Psammodynastes pulverulentus*: does the mock viper resemble a protoelapid? *J. Herpetol.* 30(1): 128–131.

JACOBSON, E. 1936. Reuzenslangen en hun prooi. *Trop. Natuur* Jub. Uitg.: 49–51.

JACOBSON, E. 1937. A case of snake-bite (*Maticora intestinalis*). *Bull. Raffles Mus.* 13: 77–79.

JACOBSON, E. 1938. *Naja bungarus* Schlegel. *Trop. Natuur* 27: 94–95.

JAEKEL, O. 1911. Die Fossilen Schildkrötenreste von Trinil. *In*: Die *Pithecanthropus*—Schichten auf Java. pp: 75–81; Pls. XIV–XV. L. Selenka & M. Blanckenhorn (Eds.). Engelmann, Leipzig.

JAN, G. 1858. Plan d'une iconographie descriptive des ophidiens, et description sommaire de nouvelles espèces de serpents. *Rev. & Mag. Zool.* 10: 514–527.

JAN, G. 1859. Plan d'une iconographie der ophidiens et description sommaire de nouvelles espèces des serpents. *Rev. & Mag. Zool.* 2(11): 122–130.

JAN, G. 1862. Sulla famiglia dei Tiflopidi sui loro generi e sulle specie del genere *Stenostoma*. *Arch. Zool. Anat. Fis., Genova* 1: 178–199.

JAN, G. & F. SORDELLI. 1864. Iconographie générale des ophidiens. Tome Premier. Livrais 1 y 17. J. B. Baillère et fils, Milan and Paris. 6 pl.

JAN, G. & F. SORDELLI. 1864. Iconographie générale des ophidiens. Premiere famille. Les Typhlopiens. J. B. Baillère et fils, Milan and Paris. 42 pp.

JENKINS, M. D. 1995. Species in danger. Tortoises and freshwater turtles: the trade in southeast Asia. Traffic International, Cambridge. iv + 48 pp.

JOGER, U. 1991. A molecular phylogeny of agamid lizards. *Copeia* 1991(3): 616–622.

JUDD, H. L., J. P. BACON, D. RÜEDI, J. GIRARD & K. BENIRSCHKE. 1977. Determination of sex in the Komodo dragon. *Int. Zoo Yrbk* 1977: 208–219.

KAJIHARA, T. & T. UCHIDA. 1974. The ecology and fisheries of the hawksbill turtle, *Eretmochelys imbricata*, in southeast Asia. *Japanese J. Herpetol.* 5(3): 48–56.

KAR, C. S. & S. BHASKAR. 1982. Status of sea turtles in the eastern Indian Ocean. *In*: Biology and conservation of sea turtles. pp: 365–372. K. A. Bjorndal (Ed.). Smithsonian Institution Press, Washington, D.C.

KAWAMURA, Y., H. CHINZEI & Y. SAWAI. 1975. Snakebites in Indonesia. *The Snake* 7(2): 73–78. [In Japanese, with English summary.]

KERN, J. A. 1968. Dragon lizards of Komodo. *Natl. Geogr.* 154(6): 874–880.

KHAN, I. H. 1987. Conservation of endangered marine species in Andamans. *In*: Proc. Symp. Manage. Coastal Ecosystems & Oceanic Res. Andamans. pp: 66–70. Andaman Science Association, Port Blair.

KHARIN, V. E. 1981. [A review of sea snakes of the genus *Aipysurus* (Serpentes: Hydrophiidae).] *Zool. Zh.* 60: 257–264. [In Russian.]

KHARIN, V. E. 1984. [A new species of the genus *Hydrophis sensu lato* (Serpentes: Hydrophiidae) from the North Australian Shelf.] *Zool. Zh.* 62: 1751–1753. [In Russian.]

KHARIN, V. E. 1984. [A review of sea snakes of the group *Hydrophis* sensu lato (Serpentes, Hydrophiidae).] 3. The genus *Leioselasma*. *Zool. Zh.* 63: 1536–1546. [In Russian.]

KHARIN, V. E. 1984. [Revision of sea snakes of subfamily Laticaudinae Cope, 1879, sensu lato (Serpentes, Hydrophiidae).] *Proc. Zool. Inst., Leniingrad* 124: 128–139. [In Russian.]

KHARIN, V. E. 1989. [A new sea snake species of the genus *Disteira* (Serpentes, Hydrophiidae) from the waters of the Malay Archipelago.] *Vêst. Zool.* 1989: 29–32. [In Russian.]

KICHENER, H. J. 1953. Skin change of the flying gecko (*Ptychozoon kuhli*), Chichak kubin. *Malay. nat. J.* 7: 180–181.

KICHENER, H. J. 1955. A gecko colony. *Malayan nat. J.* 10: 15–16.

KIEW, B. H. 1987. The flying gecko—*Ptychozoon*. *Nat. Malaysiana* 12(1): 18–19.

KIEW, B. H. 1984. *Rana malesiana*, a new frog from the Sunda region. *Malayan nat. J.* 37: 153–161.

KIEW, B. H. 1984. A new species of burrowing frog (*Calluella flava* sp. nov.) from Borneo. *Malayan nat. J.* 37: 163–166.

KIEW, B. H. 1984. A new species of frog (*Kalophrynus baluensis* n. sp.) from Mount Kinabalu, Sabah, Malaysia. *Malayan nat. J.* 38: 151–156.

KIEW, B. H. 1984. Conservation status of the Malaysian fauna. IV. Turtles, terrapins and tortoises. *Malayan Natural.* 38(2): 2–3.

KING, D. 1991. The effect of body size on the ecology of varanid lizards. *Mertensiella* 2: 204–210.

KING, F. W. 1968. Ora-giants of Komodo. *Anim. Kingdom* August: 2–9.

KING, F. W. 1969. The giant lizards of Komodo. *Nat. & Sci.* 7(1): 5–7.

KING, F. W. 1978. A new Bornean lizard of the genus *Harpesaurus*. *Sarawak Mus. J.* 26: 205–209.

KING, F. W. & R. L. BURKE. 1989. Crocodilian, tuatara, and turtle species of the world: A taxonomic and geographic reference. Association of Systematics Collections, Washington, D.C. xxii + 216 pp.

KING, F. W. & R. F. INGER. 1961. A new cave-dwelling lizard of the genus *Cyrtodactylus* from Niah. *Sarawak Mus. J.* 10: 274–276.

KIRSCHNER, A. 1997. Anmerkungen zu *Python curtus curtus*. *Elaphe* 5(2): 11–12.

KITCHELL, K. & H. A. DUNDEE. 1994. Translation and annotation of the amphibian and reptile section of Systema Naturae X. *In*: A trilogy of the herpetology of Linnaeus's Systema Naturae. pp: 3–40. *Smithsonian Herpetol. Inf. Serv.* (100): 1–61.

KITCHING, R. L. & A. G. ORR. 1996. The foodweb from water-filled tree-holes in Kuala Belalong, Brunei. *Raffles Bull. Zool.* 44(2): 405–413.

KLEMMER, K. 1963. Liste der rezenten Giftschlangen. Elapidae, Hydrophiidae, Viperidae und Crotalidae. *In*: Gifischlangen der Erde. Sondeband, Behringwerke-Metteilunge, Marburg/Lahn: 253–464.

KLINGEL, H. 1965. Über das Flugverhalten von *Draco volans* (Agamidae) und verwandten Arten. *Zool. Anz.* 175: 273–281.

KLUGE, A. G. 1968. Phylogenetic relationships of the gekkonid lizard genera *Lepidodactylus* Fitzinger, *Hemiphyllodactylus* Bleeker, and *Pseudogekko* Taylor. *Philippines J. Sci.* 95: 331–452.

KLUGE, A. G. 1985. Notes on gekko nomenclature (Sauria: Gekkonidae). *Zool. Med. Leiden* 59(10): 95–100.

KLUGE, A. G. 1991. Checklist of gekkonoid lizards. *Smithsonian Herpetological Inf. Serv.* (85): 1–35.

KLUGE, A. G. 1993. Gekkonoid lizard taxonomy. International Gecko Society, San Diego. 245 pp.

KLUGE, A. 1997. A review of varanid lizard phylogeny: old hypothesis versus new data. *In*: Herpetology '97. Abstracts of the Third World Congress of Herpetology. 2–10 August, 1997. pp: 114. Z. Rocek & S. Hart (Eds). Third World Congress of Herpetology, Prague.

KOCK, W. 1859. Slangen van Buitenzorg. *Natuur. Tijd. Ned-Indië* 16(2): 207–208.

KOPSTEIN, P. F. 1926. Reptilien von den Molukken und den Benachbarten Inseln. *Zool. Med.* 1: 71–112.

KOPSTEIN, F. 1927. Über einen fall von polymorphismus bei der Scincidengattung *Lygosoma*. *Treubia* 14(4): 371–375; Pl. 9.

KOPSTEIN, F. 1927. Die reptilienfauna der Sula-Inseln. *Treubia* 14(4): 437–446.

KOPSTEIN, F. 1927. Over het verslinden van menschen door *Python reticulatus*. *Trop. Natuur* 16: 65–67.

KOPSTEIN, F. 1928. De javaansche brilslang *Naja tripudians sputatrix*. *Trop. Natuur* 17: 191–199.

KOPSTEIN, F. 1929. Observations on the venomous effect of *Naja bungarus*. *Med. Dienst. Volks. Ned.-Indië*. 1929: 1–6.

KOPSTEIN, F. 1929. Herpetologische Notizen I. Ein neuer Fall von Termitophillie. *Treubia* 10: 467–469.

KOPSTEIN, F. 1930. Herpetologische Notizen II. Oologische Beobachtungen an West-Javanischen Reptilien. *Treubia* 11: 301–307.

KOPSTEIN, F. 1930. Herpetologische Notizen III. Reptilien des (tm)stlichen Preanger (West-Java). *Treubia* 12(3/4): 273–276.

KOPSTEIN, F. 1930. Die Giftschlangen Javas und ihre Bedeutung für den Menschen. *Z. Morph. (tm)kol. Tiere* 19: 339–363.

KOPSTEIN, F. 1931. Herpetologische Notizen IV. *Fordonia leucobalia* Schlegel und *Cerberus rhynchops* Schneider. *Treubia* 13: 1–4.

KOPSTEIN, F. 1932. Herpetologische Notizen V. *Bungarus javanicus*, eine neue giftschlange von Java. VI. Weitere Beobachtungen über die fortpflanzung West-Javanischer Reptilien. *Treubia* 14(1): 78–84.

KOPSTEIN, F. 1935. Termitophilie bei javanischen Reptilien. *Ent. Med. Ned.-Indië* 1: 54–56.

KOPSTEIN, F. 1935. Herpetologische Notizen VII–X. Reptilien vom Karimata-Archipel; Reptilien von der Insel Boeton; Reptilien von Benkoelen; Weitere Beobachtungen über die Fortpflanzung west-javanischer Reptilien. *Treubia* 15(1): 51–56.

KOPSTEIN, F. 1935. Herpetologische Notizen XI. Bemerkungen *Naja naja leucodira* (Blgr.). *Treubia* 15: 151–155.

KOPSTEIN, F. 1936. Herpetologische Notizen XII–XV. Ein Beitrag zur Herpetofauna von Celebes. *Treubia* 15(3): 255–266.

KOPSTEIN, F. 1936. Ueber *Bungarus javanicus* Kopst. *Treubia* 15(3): 265–266.

KOPSTEIN, F. 1937. Schlangen von Enggano. *Treubia* 16: 239–244.

KOPSTEIN, F. 1938. Herpetologische Notizen XVIII. Über die systematische Stellung der Art *Trimeresurus gramineus* (Shaw) von den Sunda Inseln. *Treubia* 16: 325–333.

KOPSTEIN, F. 1938. Ueber pigmentierungsanomalien bei Malaiischen Reptilien. *Treubia* 16: 361–366.

KOPSTEIN, F. 1938. Ein Beitrag zur Eierkunde und zur Fortpflanzung der malaiischen Reptilien. *Bull. Raffles Mus.* 14: 81–167.

KOPSTEIN, F. 1938. Ein Beitrag zur Morphologie, Biologie und Okologie von *Xenodermus javanicus* Reinhardt. *Bull. Raffles Mus.* 14: 168–174.

KOPSTEIN, F. 1941. Über sexualdimorphismus bei malaiischen Schlangen. *Temminckia* 6: 109–185.

KOSUCH, J., M. VEITH & A. SEITZ. 1994. Genetische Differenzierungssmuster verschiedener Fröscharten (genus *Rana* s.l.) auf den Großen Sundainseln (Java, Sumatra, Borneo: Indonesien). *In*: Verhandlungen der DZG, Kurzpublikationen 87. Jahresversammlung 1994. Jena.

KOSUCH, J., A. OHLER & M. VEITH. 1997. Genetic and morphological variation in *Limnonectes limnocharis* (Gravenhorst, 1829) from the Great Sunda Islands (Sumatra, Java, Borneo). *In*: Herpetology '97. Abstracts of the Third World Congress of Herpetology. 2–10 August, 1997. Z. Rocek & S. Hart (Eds). pp: 249. Third World Congress of Herpetology, Prague.

KRATZER, H. 1973. [Observations on the incubation time of a clutch of eggs of *Varanus salvator*.] *Salamandra* 9: 27–33. [In German.]

KREBS, U. 1977. Der Dumerilwaran (*Varanus dumerilii*), ein spezialisierter Krabben fresser? *Salamandra* 15(3): 146–157.

KREBS, U. 1991. Ethology and learning from observation to semi-natural experiment. *Mertensiella* 2: 220–232.

KUCH, U. 1996. Erfolgreiche Terrarienhaltung eines Java-Kraits, *Bungarus javanicus* Kopstein, 1932. *Elaphe* 4(2): 10–12.

KUCH, U. & W. SCHEYER. 1996. Erfahrungen mit der terrarienhaltung von vier arten nahrungsspezialisierter giftnattern der gattung *Bungarus* Daudin, 1803. Teil IV: *Bungarus flaviceps* (Reinhardt, 1843). *Sauria, Berlin* 18(2): 3–16.

KUHL, H. 1820. Beiträg zur Zoologei und Vergleichenden Anatomie. Erste Abth. Beitrage zur Zoologie. Frankfurt am Main. 8 + 152 pp.

KUHL, H. 1824. Sur les reptiles de Java. Extrait d'une lettre adressée de

Java en Hollande, par M. Kuhl, datée de Pjihorjavor au pied du Pangerango, le 18 Juillet 1821. *Bull. Univ. Sci. Indust. Férussac (Ser. 2 Bull. Sci. nat. Geol.)* 2: 79–83.

KUHL, H. 1824. Second lettre de M. Kuhl sur les reptiles de l'île de Java. *Bull. Univ. Sci. Industries Férussac (Ser. 2 Bull. Sci. nat. Géol.)* 2: 370–371.

KUHL, H. & J. C. VAN HASSELT. 1822. Uittreksels uit brievenvan de heeren Kuhl en van Hasselt, aan de heeren C. J. Temminck, Th. van Swinderen W. de Haan, geschreven in de Straat Sunda, den 17 den December 1820 [parts 1–3]. *Alegemeene Konst-en Letter-Bode, Haarlem* 7: 82–88, 99–104, 149–153.

KUHL, H. & J. C. VAN HASSELT. 1822. Aus einem Schreiben von Dr. Kuhl und Dr. van Hasselt auf Java. An Professor Th. van Swinderen zu Gröningen. *Isis von Oken* [Vol. I—1822] 10(4): columns 472–476.

LACEPÈDE, B. G. E. L. COMPTE DE. 1789. Histoire naturelle des quadrupèdes ovipares et des serpens. Tome 2. Imprimerie du Roi, Hôtel de Thou, Paris. 19 + 144 + 527 pp; 22 pl.

LACEPÈDE, B. G. E. L. COMPTE DE. 1804. Mémoire sur plusieurs animaux de la Nouvelle Hollande dont la description n'pas encore été publiée. *Ann. Mus. Nat. Hist. nat. Paris* 4: 184–211.

LADIGES, W. 1939. Herpetologische Beobachtungen auf Sumatra. *Zool. Anz.* 128(3/4): 235–249.

LADING, E. & R. STUEBING. 1997. Nest of a false gharial from Sarawak. *Croc. Spec. Gr. Newsl.* 16(2): 12–13.

LADING, E. A., R. STUEBING & H. K. VORIS. 1991. A population size estimate of the yellow-lipped sea krait, *Laticauda colubrina*, on Kalampunian Damit Island, Sabah, Malaysia. *Copeia* 1991(4): 1139–1142.

LALLEMONT, G. F. M. A. 1929. De Reuzen Leguaan (*Varanus komodoensis*) op de eilanden Komodo en Rintja. *Trop. Natuur* 18(8): 125–128.

LAL MOHAN, R. S. 1983. Marine turtle resources. *Bull. Cent. Mar. Fish. Res. Inst.* 34: 98–99.

LAL MOHAN, R. S. 1983. Saltwater crocodile resources. *Bull. Cent. Mar. Fish. Res. Inst.* 34: 102–103.

LANGE, J. 1989. Observations on the Komodo monitor, *Varanus komodoensis* in the Zoo-Aquarium Berlin. *Int. Zoo Yrbk* 28: 151–153.

LANZA, B. 1952. Su un esemplare di *Pelamis platyurus* (L.) catturato in un fiume della piccola Andaman (Serpentes; Hydrophiidae). *Monitore Zoologico Italiano* 62(2): 67–70.

LAURENT, R. F. 1943. Sur l'ostéologie de deux Ranides exotiques. *Bull. Mus. Hist. Nat. Belgique, Bruxelles* 19(27): 1–4.

LAURENT, R. F. 1943. Contribution a l'ostéologie et a la systématique des Rhacophorides non-Africains. *Bull. Mus. Hist. Nat. Belgique, Bruxelles* 19(28): 1–16.

LAURENT, R. F. 1948. Notes sur quelques reptiles appartenant a la collection du Musée royal d'Histoire Naturelle de Belgique. II. Formes asiatiques et néoguinéennes. *Bull. Mus. Hist. Nat. Belgique, Bruxelles* 24 (17): 1–12.

LAURENTI, J. N. 1768. Specimen Medicum exhibens Synopsin Reptilium emendatam cum experimentis circa venena et antidota reptilium austriacorum. Joan. Thomæ and Trattnern, Viennæ (= Vienna). (6) + 214 + (3) pp + 5 pl. [The real author of the work is apparently Winterl: see Fejérváry, 1917; Leydig, 1872; Stejneger, 1936.]

LAZELL, J. D., Jr. 1987. A new flying lizard from the Sangihe Archipelago, Indonesia. *Breviora* (488): 1-9.

LAZELL, J. D., Jr. 1987. Beyond Wallace Line. Biogeography travels of Flag 172. *Explorers J.* 65(2): 82-88.

LAZELL, J. D., Jr. 1992. New flying lizards and predictive biogeography of two Asian archipelagos. *Bull. Mus. Comp. Zool.* 152(9): 475-505.

LAZELL, J. D., Jr, J. E. KEIRANS & G. A. SAMUELSON. 1991. The Sulawesi black racer, *Coluber* (*Ptyas*) *dipsas*, and a remarkable parasitic aggregation. *Pacific Sci.* 45: 355-361.

LEDERER, G. 1942. Der Drachenwaren (*Varanus komodoensis* Ouwens). *Zool. Gart, Leipzig* n.s. 14(5/6): 227-244.

LEH, C. M. U., S. K. POON & Y. C. SIEW. 1985. Temperature-related phenomena affecting the sex of green turtle (*Chelonia mydas*) hatchlings in the Sarawak Turtle Islands. *Sarawak Mus. J.* 34(55): 183-193.

LEMEN, C. A. & H. K. VORIS. 1981. A comparison of reproductive strategies among marine snakes. *J. Anim. Ecol.* 50: 89-101.

LEVITON, A. E. 1955. Systematic notes on the Asian snake *Lycodon subcinctus*. *Philippine J. Sci.* 84: 195-203.

LEVITON, A. E. 1963. Remarks on the zoogeography of Philippine terrestrial snakes. *Proc. California Acad. Sci.* 31(15): 369-416.

LEVITON, A. E. 1968. The venomous terrestrial snakes of East Asia, India, Malay, and Indonesia. *In*: Venomous animals and their venoms. 1: 529-576. E. Bucherl, E. Buckley & V. Deulofeu (Eds.). Academic Press, New York.

LEVITON, A. E. & W. C. BROWN. 1959. A review of the snakes of the genus *Pseudorabdion* with remarks on the status of the genera *Agrophis* and *Typhlogeophis*. *Proc. Calif. Acad. Sci.* (4): 29: 475-508.

LEYDIG, F. 1872. Die in Deutschland lebenden Arten der Saurier. H. Laupp'schen Buchhandlung, Tübingen. 262 pp; 12 pl.

LIE, G. L. 1958. Sedikit tentang daerah disekitar Gersik. *Penggemar Alam* 37(3-4): 67-78.

LIEM, (D.) S. S. 1970. The morphology, systematics and evolution of the Old World treefrogs (Rhacophoridae and Hyperoliidae). *Fieldiana Zool.* 57: 1-145.

LIEM, D. S. S. 1971. The frogs and toads of Tjibodas National Park, Mt. Gede, Java, Indonesia. *Philippines J. Sci.* 100(2): 131-161; Pl. 1-2.

LIEM, K. F. 1959. The breeding habits and development of *Rana chalconota* (Schlegel)(Amphibia). *Treubia* 25(1): 89-111.

LIEM, K. F. 1961. On the taxonomic status and the granular patches of the Javanese frog *Rana chalconota* Schlegel. *Herpetologica* 17(1): 69-71.

LIEW, H.-C., E.-H. CHAN, F. PAPI & P. LUTSCHI. 1996. Long distance migration of green turtles from Redang Island, Malaysia: the need for regional cooperation in sea turtle conservation. *In*: International Congress of Chelonian Conservation. Proceedings. pp: 73–75. SOPTOM (Ed.). Editions Soptom, Gonfaron.

LILLEY, G. 1989. Notes on breeding the Komodo Dragon. *The Herptile* 14(4): 167–168.

LILLEY, R. 1996. Amphibians. *In*: Indonesian heritage. Volume 5. pp: 30–31. Wildlife. T. Whitten & J. Whitten (Eds). Editions Didier Millet/Archipelago Press, Singapore.

LIM, B. L. 1964. Notes on the elephant's trunk snake and the puff-faced water snake. *Malayan nat. J.* 1964: 179–183.

LIM, B. L. 1967. Further comments on rare snakes. *Fed. Mus. J.* n.s. 12: 123-126.

LIM, B. L. 1971. Venomous snakes of southeast Asia. *Southeast Asian J. Trop. Med. & Publ. Health* 2(1): 56–64.

LIM, B. L. 1990. Venomous land snakes of Malaysia. *In*: Snakes of medical importance (Asia-Pacific region). pp: 387–417. P. Gopalakrishnakone & L. M. Chou (Eds.). National University of Singapore, Singapore.

LIM, F. L. K. & M. T.-M. LEE 1989. Fascinating snakes of Southeast Asia— An introduction. Tropical Press Sdn. Bhd., Kuala Lumpur. xviii + 124 pp.

LINCOLN, G. A. 1974. Predation of incubator birds (*Megapodius freycineti*) by Komodo dragons (*Varanus komodoensis*). *J. Zool., London* 174: 419–428.

LINDHOLM, W. A. 1929. Revidiertes Verzeichnis der Gattung der rezenten Schildkröten nebst Notizen zur Nomenclatur einiger Arten. *Zool. Anz.* 81(11–12): 275–295.

LINDSAY, C. 1995. Turtle Islands: Balinese ritual and the green turtle. Takarajima Books, New York. 124 pp.

LINNAEUS, C. 1758. Systema naturae per regna tria naturae, secudum classes, ordines, genera, species, cum characteribus, differentiis, synonymis, locis. Tomus I. Editio Decima, Reformata. Laurentii Salvii, Stockholm. (4) + 823 + (1) pp. Reprinted 1894 by Deutsch Zoologische Gesselschaft and 1939 and 1956 by the British Museum (Natural History). 824 pp. + Emendanda and Addenda.

LINNAEUS, C. 1766. Systema naturae per regna tria naturae, secundum classes, ordines, genera, species, cum characteribus, differentiis, synonymis, locis. Tomus I. Editio Duodecima, Reformata. Laurentii Salvii, Stockholm. (4) + 823 + (1) pp. Section on Amphibia (Tomus I, Pars II, 1767: pp: 347–393, addenda [unpaginated in the original], reprinted 1963 by the Ohio Herpetological Society.

LINNAEUS, C. 1771. Mantissa Plantarum altera generum editionis VI. & Speciesrum editionis II. Part 2. pp: 143–584 + 10 unnumbered pages. Page 528 reprinted 1963 by the Ohio Herpetological Society.

LLOYD, M. R., R. F. INGER & F. W. KING. 1968. On the diversity of reptile and amphibian species in a Bornean rain forest. *Amer. Natural.* 102(928): 497–515.

54   DAS

LONNBERG, E. 1899. On a small collection of Javanese reptiles containing a new species of snake. *Zool. Anz.* 22(580): 108–111.

LONNBERG, E. 1928. Notes on *Varanus komodoensis* Ouwens and its affinities. *Ark. Zool.* 19A(27): 1–11.

LONNBERG, E. & H. RENDAHL. 1925. Dr. E. Mjöberg's zoological collections from Sumatra. 2. Reptiles and batrachians. *Archiv. Zool.* 17A(23): 1–3.

LORENZ, W. 1984. Die asiatischen Schildkröten der Familie Emydidae: Pt. 2. Die Gaffungen *Cyclemys* Bell 1834 und *Notochelys* Gray 1863. *Schildkröte* 6(1): 4–20.

LOSOS, J. B. & H. W. GREENE. 1988. Ecological and evolutionary implications of diet in monitor lizards. *Biol. J. Linn. Soc.* 35: 379–407.

LOVERIDGE, A. 1938. New snakes of the genera *Calamaria*, *Bungarus* and *Trimeresurus* from Mount Kinabalu, North Borneo. *Proc. Biol. Soc. Washington* 51: 43–46.

LOVERIDGE, A. 1944. A new elapid snake of the genus *Maticora* from Sarawak, Borneo. *Proc. Biol. Soc. Washington* 57: 105–106.

LOVERIDGE, A. 1944. Errata (A new elapid snake of the genus *Maticora* from Sarawak, Borneo). *Proc. Biol. Soc. Washington* 57: VI.

LOVERIDGE, A. 1945. Reptiles of the Pacific World. The Macmillan Company, New York. xii + 259 pp.

LUDEKING, E. W. A. 1859. Reptiliën van Agam. *Natuur. Tijd. Ned-Indië* 16(2): 26–27.

LUDEKING, E. W. A. 1860. Reptiliën uit de omstreken van Fort de Kock. *Natuur. Tijd. Ned-Indië* 21(1): 437–438.

LUI, W. 1996. Captive husbandry and breeding of the Malaysian cat geckos, *Aeluroscalabotes felinus*. *Reptilian* 4(3): 48–53.

LUTZ, D. & J. M. LUTZ. 1991. Komodo. The living dragon. DIMI Press, Salem, Oregon. xv + 174 pp.

LUXMOORE, R. 1996. The trade in reptiles. *In*: Indonesian heritage. Volume 5. pp: 106–107. Wildlife. T. Whitten & J. Whitten (Eds). Editions Didier Millet/Archipelago Press, Singapore.

LUXMOORE, R. & B. GROOMBRIDGE. 1990. Asian monitor lizards. A review of distribution, status, and exploitation and trade in four selected species. World Conservation Monitoring Centre, Cambridge. 195 pp.

LYONS, D. D. 1969. Keys to shell-bones of Bornean emydid turtles. *Sarawak Mus. J.* 17: 89–95.

McCANN, C. 1953. Distribution of the Gekkonidae in the Pacific area. *Proc. 7th Pacific Sci. Congr.* 4 (Zool.): 27–32.

McCARTHY, C. J. 1986. Relationships of the laticaudine sea snakes (Serpentes: Elapidae: Laticaudinae). *Bull. British Mus. nat. Hist. (Zool.)* 50: 127–161.

McCARTHY, C. J. 1993. *Laticauda* Laurenti, 1768. *In*: Endoglyphs and other major venomous snakes of the world: A checklist. pp: 145–148. P. Golay, H. M. Smith, D. G. Broadley, J. R. Dixon, C. McCarthy, J.-C. Rage, B. Schätti & M. Toriba (Eds.). Azemiops S. A., Herpetological Data Centre, Geneva. 478 pp.

McCARTHY, C. J. 1993. *Astrotia* Fischer, 1855. *In:* Endoglyphs and other major venomous snakes of the world: A checklist. pp: 224–225. P. Golay, H. M. Smith, D. G. Broadley, J. R. Dixon, C. McCarthy, J.-C. Rage, B. Schätti & M. Toriba (Eds.). Azemiops S. A., Herpetological Data Centre, Geneva. 478 pp.

McCARTHY, C. J. 1993. *Disteira* Lacépède, 1804. *In:* Endoglyphs and other major venomous snakes of the world: A checklist. pp: 225–226. P. Golay, H. M. Smith, D. G. Broadley, J. R. Dixon, C. McCarthy, J.-C. Rage, B. Schätti & M. Toriba (Eds.). Azemiops S. A., Herpetological Data Centre, Geneva. 478 pp.

McCARTHY, C. J. 1993. *Enhydrina* Gray, 1849. *In:* Endoglyphs and other major venomous snakes of the world: A checklist. pp: 227–228. P. Golay, H. M. Smith, D. G. Broadley, J. R. Dixon, C. McCarthy, J.-C. Rage, B. Schätti & M. Toriba (Eds.). Azemiops S. A., Herpetological Data Centre, Geneva. 478 pp.

McCARTHY, C. J. 1993. *Hydrophis* Latreille, 1801. *In:* Endoglyphs and other major venomous snakes of the world: A checklist. pp: 229–242. P. Golay, H. M. Smith, D. G. Broadley, J. R. Dixon, C. McCarthy, J.-C. Rage, B. Schätti & M. Toriba (Eds.). Azemiops S. A., Herpetological Data Centre, Geneva. 478 pp.

McCARTHY, C. J. 1993. *Kerilia* Gray, 1849. *In:* Endoglyphs and other major venomous snakes of the world: A checklist. pp: 242–243. P. Golay, H. M. Smith, D. G. Broadley, J. R. Dixon, C. McCarthy, J.-C. Rage, B. Schätti & M. Toriba (Eds.). Azemiops S. A., Herpetological Data Centre, Geneva. 478 pp.

McCARTHY, C. J. 1993. *Lapemis* Gray, 1834. *In:* Endoglyphs and other major venomous snakes of the world: A checklist. pp: 243–245. P. Golay, H. M. Smith, D. G. Broadley, J. R. Dixon, C. McCarthy, J.-C. Rage, B. Schätti & M. Toriba (Eds.). Azemiops S. A., Herpetological Data Centre, Geneva. 478 pp.

McCARTHY, C. J. 1993. *Pelamis* Daudin, 1803. *In:* Endoglyphs and other major venomous snakes of the world: A checklist. pp: 245–247. P. Golay, H. M. Smith, D. G. Broadley, J. R. Dixon, C. McCarthy, J.-C. Rage, B. Schätti & M. Toriba (Eds.). Azemiops S. A., Herpetological Data Centre, Geneva. 478 pp.

McCARTHY, C. J. 1993. *Thalassophina* M. A. Smith, 1926. *In:* Endoglyphs and other major venomous snakes of the world: A checklist. pp: 247–248. P. Golay, H. M. Smith, D. G. Broadley, J. R. Dixon, C. McCarthy, J.-C. Rage, B. Schätti & M. Toriba (Eds.). Azemiops S. A., Herpetological Data Centre, Geneva. 478 pp.

McCORD, W. P., J. B. IVERSON & (P.) BOEADI. 1995. A new batagurid turtle from northern Sulawesi, Indonesia. *Chelonian Conserv. & Biol.* 1(4): 311–316.

McCOSKER, J. E. 1975. Feeding behavior of Indo-Australian Hydrophiidae. *In*: The biology of sea snakes. pp: 217–232. W. A. Dunson (Ed.). University Park Press, Baltimore.

McDOWELL, S. B. Jr. 1964. Partition of the genus *Clemmys* and related problems in the taxonomy of the aquatic Testudinidae. *Proc. Zool. Soc. London* 143(2): 239–279.

McDOWELL, S. B. Jr. 1972. The genera of sea-snakes of the *Hydrophis* group (Serpentes: Elapidae). *Trans. Zool. Soc. London* 32: 189–247.

McDOWELL, S. B. Jr. & C. M. BOGERT. 1954. The systematic position of

Lanthonotus and the affinities of the anguinomorphan lizards. *Bull. Amer. Mus. nat. Hist.* 105: 1–142; 16 pl.; 1 folding fig.

MACKINNON, J. 1979. A glimmer of hope for Sulawesi. *Oryx* 15(1): 55–59.

McKINSTRY, D. M. 1983. Morphological evidence of toxic saliva in colubrid snakes: a checklist of world genera. *Herpetol. Rev.* 14(1): 12–15.

McNAB, B. K. & W. AUFFENBERG. 1976. The effect of large body size on the temperature regulation of the Komodo dragon, *Varanus komodoensis. Comp. Biochem. Physiol.* 55A: 345–350.

McNEELY, J. & P. S. WACHTEL. 1986. The cobra king. *Intl. Wildl.* 16(1): 38–41.

McNEELY, J. A. & P. S. WACHTEL. 1988. Soul of the tiger. Doubleday, New York. Reprinted 1991, Oxford University Press, Singapore. 390 pp.

MACVEIGH, W. P. 1963. Cat snake, *Boiga dendrophila* again. *Malayan nat. J.* 17(1 & 2): 61.

MAHNERT, V. 1976. Catalogue des types de poissons, amphibiens et reptiles du Muséum d'Histoire naturelle de Genève. *Revue Suisse Zool.* 83(2): 471–496.

MALEYEV, Y. 1962. The Komodo dragon. *Bull. Philadelphia Herpetol. Soc.* 10(1): 32–34.

MALEYEV, Y. & I. S. DAREVSKY. 1963. [Dragons of Komodo.] *Priroda, Moscow* (3): 24–25.

MALHOTRA, A. & R. S. THORPE. "1997" 1996. New perspectives on the evolution of south-east Asian pit vipers (genus *Trimeresurus*) from molecular studies. *In*: Venomous snakes: Ecology, evolution and snakebite. pp: 115–128. R. S. Thorpe, W. Wüster & A. Malhotra (Eds.). Symp. Zool. Soc. London (70). The Zoological Society of London/Clarendon Press, Oxford.

MALHOTRA, A. & R. S. THORPE. 1997. Too many *Trimeresurus*? New perspectives from molecular studies. *In*: Herpetology '97. Abstracts of the Third World Congress of Herpetology. 2–10 August, 1997. pp: 134. Z. Rocek & S. Hart (Eds). Third World Congress of Herpetology, Prague.

MALKMUS, R. 1985. Amphibien und Reptilien vom Mount Kinabalu (4101 m), Nordborneo. *Herpetofauna, Weinstadt* 7(35): 6–13.

MALKMUS, R. 1986. Ein seltener Ruderfrösch vom Mount Kinabalu-Nordborneo *Rhacophorus everetti macrosceles* (Boulenger). *Sauria, Berlin* 8(1): 3–4.

MALKMUS, R. 1987. Herpetologische Beobachtungen am Mt. Kinabalu, Borneo. *Mitt. Zool. Mus. Berlin* 63(2): 269–292.

MALKMUS, R. 1988. *Cyrtodactylus baluensis* (Mocquard, 1890). *Sauria, Berlin* 10(1): 29.

MALKMUS, R. 1988. *Bufo juxtasper* als bewehmer der kulturlandschaft tambunans (Nord-Borneo). *Sauria, Berlin* 10(4): 9–11.

MALKMUS, R. 1988. Herpetologische Studien an einem Waldbach am Mt. Kinabalu, Borneo. *Herpetofauna, Weinstadt* 10(53): 6–11.

MALKMUS, R. 1988. Wanderungen im Mount Kinabalu—Nationalpark/Nordborneo. *Nat. Mus., Frankfurt/M* 118(6): 161–181.

MALKMUS, R. 1989. Herpetologische Beobachtungen am Mt. Kinabalu, Borneo. II. *Mitt. Zool. Mus. Berlin* 65(2): 179–200.

MALKMUS, R. 1991. *Sphenomorphus aquaticus* Malkmus 1991 (Sauria: Scincidae) vom Mt. Kinabalu/Nordborneo. *Sauria, Berlin* 13(3): 23–28.

MALKMUS, R. 1991. Die Bogenfingergeckos der gattung *Cyrtodactylus* am Mt. Kinabalu/Nord-borneo. *Herpetofauna, Weinstadt* 13(74): 6–10.

MALKMUS, R. 1991. Zur herpetofauna des oberen Liwago/Mount Kinabalu/Nord-borneo. *Herpetofauna, Weinstadt* 13(72): 26–34.

MALKMUS, R. 1992. Herpetologische Beobachtungen am Mount Kinabalu, Nord-borneo. III. *Mitt. Zool. Mus. Berlin* 68(1): 101–138.

MALKMUS, R. 1992. *Leptolalax pictus* sp.n. (Anura: Pelobatidae) vom Mount Kinabalu/Nord-Borneo. *Sauria, Berlin* 14(3): 3–6.

MALKMUS, R. 1993. Herpetologische streitrüge in den Regenwäldern von Borneo. *DATZ* (8): 516–520.

MALKMUS, R. 1993. Die Folgen eines Bisses von *Maticora intestinalis everetti*. *Salamandra* 29(2): 153–154.

MALKMUS, R. 1993. Zur Zeichnungsvariabilität und zum Totstellverhalten des Ruderfrösches *Rhacophorus angulirostris* Mocquard, 1890. *Sauria, Berlin* 15(3): 35–38.

MALKMUS, R. 1993. Bemerkungen zu einer kleinen Sammlung von Amphibien und Reptilien aus Nordost-Sulawesi. *Mitt. Zool. Mus. Berlin* 69(1): 175–184.

MALKMUS, R. 1993. Herpetologische Streifzüge in den Regenwäldern von Borneo. *DATZ* 46(8): 516–520.

MALKMUS, R. 1994. Vijf kikkers bij de Mesilau-grot op Mt. Kinabalu (noordelijk Borneo). *Lacerta* 52(4): 86–90.

MALKMUS, R. 1994. Herpetologische Beobachtungen am Mt. Kinabalu, Nord-Borneo. IV. *Mitt. Zool. Mus. Berlin* 70(2): 101–138.

MALKMUS, R. 1994. Een zeldzame slang van Mt. Kinabalu: Schmidts grondslang (*Calamaria schmidti*). *Lacerta* 52(5): 111–113.

MALKMUS, R. 1994. Einige Bemerkungeen zu *Kalophrynus baluensis* Kiew, 1984. *Sauria, Berlin* 16(2): 9–13.

MALKMUS, R. 1995. Over het voortplantingsbiotoop van *Philautus* sp. Mt. Kinabalu (noordelijke Borneo). *Lacerta* 53(3): 68–74.

MALKMUS, R. 1995. Wer rief hier: Frösch oder Heuschrecke? *Sauria, Berlin* 17(2): 35–38.

MALKMUS, R. 1996. Voortplanting van de Dwergpad *Pelophryne misera*. *Lacerta* 54(4): 129–133.

MALKMUS, R. 1996. Kopfseitenflecken bei beiden Geschlechtern von *Rhacophorus angulirostris* Ahl, 1927 (Amphibia: Rhacophoridae). *Salamandra* 31(4): 245–236.

MALKMUS, R. 1996. Borneose Bergagamen (*Phoxophrys*) von Mount Kinabalu. *Lacerta* 54(2): 65–69.

MALKMUS, R. & M. KUNKEL. 1992. Amphibien und Reptilien aus dem Leuser-Nationalpark/Nord-Sumatra. *Herpetofauna, Weinstadt* 14(76): 27–34.

MALKMUS, R. & K. REIDE. 1993. Nachtrag zu *Leptolalax pictus* Malkmus, 1992 (Anura: Pelobatidae). *Sauria, Berlin* 15(3): 7–9.

MALKMUS, R. & K. REIDE. 1996. Die Baumfrösche der Gattung *Philautus* vom Mount Kinabalu—Teil I: Überblick und die *aurifasciatus*-Gruppe mit Beschreibung einer neuen Art (*Philautus saueri* n. sp.). *Sauria, Berlin* 18(1): 27–37.

MALKMUS, R. & K. REIDE. 1996. Die Baumfrösche der Gattung *Philautus* vom Mount Kinabalu—Teil II: Die *vermiculatus*-Gruppe mit Beschreibung einer neuen Unterart (*Philautus aurantium gunungensis* n. ssp.) und die *hosei*-Gruppe. *Sauria, Berlin* 18(2): 21–28.

MALKMUS, R. & K. REIDE. 1996. Zur Bioakustik von *Megophrys baluensis* (Boulenger, 1899) und *Kalophrynus baluensis* Kiew, 1984, zweier seltener Froscharten aus Borneo. *Herpetozoa* 9(1/2): 151–155.

MALKMUS, R. & H. SAUER. 1993. Die Folgen eines Bisses von *Maticora intestinalis everetti*. *Salamandra* 29(2): 153–154.

MALKMUS, R. & H. SAUER. 1996. Ruhestellung von *Pareas nuchalis* und Erstnachweis dieser Art im Nationalpark Mount Kinabalu/Malaysia. *Salamandra* 32(1): 55–58.

MALNATE, E. V. 1960. Systematic division and evolution of the colubrid snake genus *Natrix*, with comments on the subfamily Natricinae. *Proc. Acad. nat. Sci. Philadelphia* 112(3): 41–71.

MALNATE, E. V. 1971. A catalog of primary types in the herpetological collections of the Academy of Natural Sciences, Philadelphia (ANSP). *Proc. Acad. Nat. Sci. Philadelphia* 123(9): 345–375.

MALNATE, E. V. 1975. The classification and relationships of Australasian watersnakes. *American Phil. Soc. Yrbk.* 1974: 357–358.

MALNATE, E. V. & G. UNDERWOOD. 1988. Australasian natricine snakes of the genus *Tropidonophis*. *Proc. Acad. nat. Sci. Philadelphia* 140(1): 59–201.

MANAÇAS, S. 1956. Dois saurios de Timor Portugues. *Anais Jta Invest. Ultramas* 11(3): 271–277.

MANDAL, A. K. & K. N. NAIR. 1973. A new coccodium from a tree-lizard *Gymnodactylus rabitus* [sic] (Blyth) (Reptilia: Lacertilia) from Andaman Islands. *Sci. & Cult.* 39(8): 369–370.

MANSUKHANI, M. R. & A. K. SARKAR. 1980. On a new species of toad from Camorta, Andaman and Nicobar, India. *Bull. Zool. Surv. India* 3: 97–101.

MANTHEY, U. 1982. Exkursion am Mt. Kinabalu 4100 m, Nordborneo. Teil 1: Klimatische Untersuahungen im Hinblick auf die Terrarienhaltung. *Herpetofauna, Weinstadt* 4(20): 28–33.

MANTHEY, U. 1985. *Leptobrachium hendricksonii* Taylor. *Sauria, Berlin* 7(4): 1–2.

MANTHEY, U. 1990. *Harpesaurus beccarii* Doria. *Sauria, Berlin* 12(3): 1–2.

MANTHEY, U. 1991. *Gonocephalus denzeri* sp. n. (Sauria: Agamidae) aus Sarawak, Borneo. *Salamandra* 27(3): 152–157.

MANTHEY, U. 1994. *Rana hosii* Boulenger. *Sauria, Berlin* Suppl. 16(3): 319–320.

MANTHEY, U. & W. DENZER. 1982. Exkursion am Mt. Kinabalu 4100 m, Nordborneo. Teil 2: Herpetologische Eindrucke. *Herpetofauna, Weinstadt* 4(21): 11–19.

MANTHEY, U. & W. DENZER. 1982. Exkursion am Mt. Kinabalu 4100 m, Nordborneo. Teil 3. Checklist der Herpetofauna oberhalb 600m ü. NN. *Herpetofauna, Weinstadt* 5(23): 20–31.

MANTHEY, U. & W. DENZER. 1991. Die Echten Winkelkopfagamen der Gattung *Gonocephalus* Kaup (Sauria: Agamidae). I. Die *megalepis*-Gruppe mit *Gonocephalus lacunosus* sp. n. aus Nord-Sumatra. *Sauria, Berlin* 13(1): 3–10.

MANTHEY, U. & W. DENZER. 1991. Die Echten Winkelkopfagamen der Gattung *Gonocephalus* Kaup (Sauria: Agamidae). II. Allgemeine Angaben zur Biologie und Terraristik. *Sauria, Berlin* 13(2): 19–22.

MANTHEY, U. & W. DENZER. 1991. Die Echten Winkelkopfagamen der Gattung *Gonocephalus* Kaup (Sauria: Agamidae). III. *Gonocephalus grandis* (Gray, 1845). *Sauria, Berlin* 13(3): 3–10.

MANTHEY, U. & W. DENZER. 1992. Die Echten Winkelkopfagamen der Gattung *Gonocephalus* Kaup (Sauria: Agamidae). IV. *Gonocephalus mjoebergi* Smith, 1925 und *Gonocephalus robinsoni* Boulenger, 1908. *Sauria, Berlin* 14(1): 15–19.

MANTHEY, U. & W. DENZER. 1992. Die Echten Winkelkopfagamen der Gattung *Gonocephalus* Kaup (Sauria: Agamidae). V. Die *bellii*-Gruppe. *Sauria, Berlin* 14(3).

MANTHEY, U. & W. DENZER. 1993. Die Echten Winkelkopfagamen der Gattung *Gonocephalus* Kaup (Sauria: Agamidae). VI. Die *chamaeleontinus*-Gruppe. *Sauria, Berlin* 14(1): 23–28.

MANTHEY, U. & W. GROSSMANN. 1994. Gleitfluegen bei *Bronchocela cristatella* (Kuhl, 1820)(Sauria: Agamidae). *Sauria, Berlin* 16(3): 37–38.

MANTHEY, U. & W. GROSSMANN. 1997. Amphibien & Reptilien Sudostasiens. Natur und Tier, Münster. 512 pp.

MARCELLINI, D. L. 1977. The function of a vocal display of the lizard *Hemidactylus frenatus* (Sauria: Gekkonidae). *Anim. Behav.* 25: 414–417.

MARX, H. 1958. Catalogue of type specimens of reptiles and amphibians in the Chicago Museum of Natural History. *Fieldiana: Zool.* 36: 407–496.

MARX, H., J. S. ASHE & L. E. WATROUS. 1988. Phylogeny of the viperine snakes (Viperinae). Part I. Character analysis. *Fieldiana Zool. n.s.* (51): 1–16.

MARX, H. & R. F. INGER. 1955. Notes on snakes of the genus *Calamaria*. *Fieldiana Zool.* 37: 167–209.

MARX, H. & G. B. RABB. 1965. Relationships and zoogeography of the viperine snakes (Family: Viperidae). *Fieldiana Zool.* 44(21): 161–206.

MARX, K. W. 1965. A substitute name, *Edwardtayloria*, for a genus of tree frogs from southeast Asia (Anura: Rhacophoridae). *Sci. Publ. Sci. Mus. St. Paul* 2: 1–3.

MASLIN, T. P. 1942. Evidence for the separation of the crotalid genera *Trimeresurus* and *Bothrops,* with a key to the genus *Trimeresurus*. *Copeia* 1942(1): 18–24.

MATSUI, M. 1983. Amphibians from Sabah. II. Acoustic characteristics of three common anuran species. *Contrib. Biol. Lab. Kyoto Univ.* 26: 123–129.

MATSUI, M. 1986. Three new species of *Amolops* from Borneo (Amphibia: Anura, Ranidae). *Copeia* 1986(3): 623–630.

MATSUI, M. 1996. Call characteristics and systematic relationships of a Malayan treefrog *Nyctixalus pictus* (Anura, Rhacophoridae). *Herpetol. J.* 6(2): 62–64.

MATSUI, M., T. CHAN-ARD, J. NABHITABHATA. 1996. Distinct specific status of *Kalophrynus pleurostigma interlineatus* (Anura, Microhylidae). *Copeia* 1996(2); 440–445.

MATSUI, M., T. HIKIDA & H. NAMBU. 1985. The amphibians and reptiles collected from Borneo, Malaysia. *Bull. Toyama Sci. Mus.* (8): 151–159.

MATSUI, M., T. HIKIDA & H. OTA. 1984. A small collection of amphibians and reptiles from Padang, Sumatra. *In*: Forest ecology and flora of Gunung Gadut, West Sumatra (1980–1984). Sumatrana Nat. Study (Botany). M. Hotta (Ed.). 1984: 121–123.

MATSUI, M., T. SETO & T. UTSUNOMIYA. 1986. Acoustic and karyotypic evidence for specific separation of *Polypedates megacephalus* from *P. leucomystax*. *J. Herpetol.* 20(4): 483–489.

MATSUI, M., H.-S. YONG, K. ARAYA & A. A. HAMID. 1996. Acoustic characteristics and systematic relationships of arboreal microhylid frogs of the genus *Metaphrynella* from Malaysia. *J. Herpetol.* 30(3): 424–427.

MAY, R. M. 1980. Why are there fewer frogs and lizards in southeast Asia than in Central America? *Nature, London* 287: 105.

MEDWAY, LORD. 1975. The function of the membranes of *Ptychozoon* spp. *Malay. nat. J.* 29(1): 28–30.

MEHRTENS, J. M. 1970. *Orlitia* the Bornean terrapin. *Int. Turtle & Tortoise Soc. J.* 4(5): 6–7, 33.

MEHTA, H. S. & G. C. RAO. 1987. Microhylid frogs of Andaman and Nicobar Islands. *J. Andaman Sci. Assoc.* 3: 98–104.

MEIJER, A. B. 1870. Over de giftklieren *Callophis intestinalis* en *C. bivirgatus*. *Natuur. Tijd. Ned-Indië* 31(1): 223–225.

MEIJER, E. F. 1859. Reptiliën enz. van Bintang. *Tijd. Ned-Indië* 16(2): 16.

MEIJER, E. F. 1859. Over vischoorten en reptilien van Bintang en Siak. *Natuur. Tijd. Ned-Indië* 26(1): 86–88.

MEINERTZ, T. 1952. Das hertz und die grossen Blutgefasse ser Komodoedechse, *Placovaranus komodoensis* (Ouwens). *Z. Anat. Entwicklungsgesch, Berlin* 116: 315–319.

MEISE, W. & W. HENNIG. 1932. Die Schlangengattung *Dendrophis*. *Zool. Anz.* 99(11/12): 273–297.

MEISE, W. & W. HENNIG. 1935. Zur Kenntnis von *Dendrophis* und *Chrysopelea*. Ein Beitrag zur Systematischen Bewertung der Ophisthoglypha. *Zool. Anz.* 109(5–6): 138–150.

MERTENS, R. "1920" 1921. Über das im Senckenbergischen Museum befindliche Exemplar von *Cophotis sumatrana* Hubrecht (Rept., Lac.). *Senckenbergiana* (6): 179–180.

MERTENS, R. 1924. Ueber einige Reptilien aus Borneo. *Zool. Anz.* 60: 155–159.

MERTENS, R. 1927. Herpetologische Mitteilungen XVII. *Mabuya multifasciata* Kuhl auf Bali. *Senckenbergiana* 9(5): 181–182.

MERTENS, R. 1927. Herpetologische Mitteilungen XVIII. Zur Verbreitung der *Vipera russelii* Shaw. *Senckenbergiana* 9(5): 182–184.

MERTENS, R. 1927. Neue Amphibien und Reptilien aus dem Indo-Australischen Archipel, gesammelt wahrend der Sunda-Expedition Rensch. *Senckenbergiana* 9(6): 234–242.

MERTENS, R. 1928. Herpetologische mitteilungen. XXII. Zur herpetofauna der Insel Sumba. *Senckenbergiana* 10: 227–231.

MERTENS, R. 1928. Neue Inselrassen von *Cryptoblepharis boutonii* (Desjordin). *Zool. Anz.* 78: 82–89.

MERTENS, R. 1928. Ueber die zoogeographische Bedeutung der Balistrasse auf Grund der Verbreitung der Amphibien und Reptilien. *Zool. Anz.* 78: 77–82.

MERTENS, R. 1929. Bemerkungen über *Typhlops florensis. Treubia* 11: 293–296.

MERTENS, R. 1929. Über eine kleine herpetologische Sammlung aus Java. *Senckenbergiana Biol.* 11(1/2): 22–33.

MERTENS, R. 1929. On rare snakes of the Senckenberg Museum. *Bull. Antivenim Inst. America, Philadelphia 3,* 86: 41.

MERTENS, R. 1929. Über einige neu oder selten eingeführte Amphibien und Reptilien aus Sudostasien, II. *Aquar-Terr.* 40(23): 168–172; Pl. 24–25.

MERTENS, R. 1929. Herpetologische Mitteilungen, XXIII–XXV.XXIV: Amphibien und Reptilien aus Atjeh (Nordsumatra), gesammelt von Herrn H. R. Rookmaaker. *Zool. Anz.* 86(3/4): 62–66.

MERTENS, R. 1929. Zur Kenntnis der *Rana microdisca* Boettger und ihrer Rassen. *Zool. Anz.* 86(3/4): 66–68.

MERTENS, R. 1930. Die Amphibien und Reptilien der Inseln Bali, Lombok, Sumbawa, und Flores. *Abh. Senckenbergiana Naturf. Ges.* (423): 117–342.

MERTENS, R. 1931. *Ablepharus boutonii* (Desjordin) und seine geographische Variation. *Zool. Jahrb. Abt. Syst.* 61: 63–210.

MERTENS, R. 1933. Weitere Mitteilungen über die Rassen von *Ablepharis boutonii* (Desjardin). I. *Zool. Anz.* 105: 92–96.

MERTENS, R. 1934. Weitere Mitteilungen über die Rassen von *Ablepharis boutonii* (Desjardin). II. *Zool. Anz.* 108: 40–43.

MERTENS, R. 1934. Amphibien und Reptilien Deutschen Limnologischen Sunda-Expedition. *Archiv. für Hydrobiol.* 1934: 695–.

MERTENS, R. 1934. Die Insel-Reptilien, ihre Ausreitung, variation und Artbildung. *Zoologica, Stuttgart* 32(6): 1–209; 6 pl.

MERTENS, R. 1934. Die Schlangengattung *Dendrelaphis* Boulenger in systematischer und zoogeographischer Beziehung. I. *Arch. Naturges., Berlin* 3(2): 187–204.

MERTENS, R. 1936. Über einige für die Insel Bali neue Reptilien. *Zool. Anz.* 115: 126–129.

MERTENS, R. 1937. Über dussere Geschlechts-Merkmale einiger Schlangen. *Senckenbergiana* 19(3/4): 169–174.

MERTENS, R. 1937. Ueber *Callagur borneoensis* (Schlegel & Müller), eine seltene Brackwasser-Schildkröte. *Aquar-Terr.* (7): 150–151.

MERTENS, R. 1942. Zwei Bemerkungen über Schildkröten Sudost-Asiens. I. Über einen bemerkenswerten Geschlechtsunterschied bei *Geoemyda spinosa* (Gray). *Senckenbergiana* 25(1/3): 41–46.

MERTENS, R. 1942. Die Familie der Warane (Varanidae). *Abh. Senckenbergiana Natur. Ges.* (462): 1–116; (465): 117–234; (466): 235–391.

MERTENS, R. 1943. Systematische und ökologische Bemerkungen über die Regenbogenschlangen, *Xenopeltis unicolor* Reinwardt. *Zool. Garten (N. F.)* 15: 213–220.

MERTENS, R. 1946. Die Warn-und Drohrenktionen der Reptilien. *Abh. Senckenbergiana Natur. Ges.* (471): 1–103.

MERTENS, R. 1946. Über den Komodo-Waran Berliner Aquarium, besonders seinen Schädel. *Senckenbergiana* 27: 153–157.

MERTENS, R. 1950. Die tiergeographische Bedeutung der Bali-Strasse: eine Richtigstellung. *Senckenbergiana* 1(1/2): 9–10.

MERTENS, R. 1954. Über die javanische Eidechse *Dendragama fruhstorferi* und die Gattung *Dendragama*. *Senckenbergiana* 34(4–6): 185–186.

MERTENS, R. 1956. Eidechsen (Reptilia) vom Karimundjawa-Archipel. *Treubia* 23(2): 253–257.

MERTENS, R. 1957. Zur herpetofauna von Ostjava und Bali. *Senckenbergiana* 38(1/2): 23–31.

MERTENS, R. 1957. Amphibien und reptilien aus dem äusserten western Javas und von benachbarten Eilanden. *Treubia* 24(1): 83–106.

MERTENS, R. 1959. Amphibien und Reptilien von Karimundjawa, Bawean und den Kangean-Inseln. *Treubia* 25(1): 1–15.

MERTENS, R. 1959. Eine Panzerschleichse (*Ophisaurus*) aus Sumatra. *Senckenbergiana Biol.* 40(1/2): 109–111.

MERTENS, R. 1959. Liste der Warane Asiens und der Indo-australischen Inselwelt mit systematischen Bemerkungen. *Senckenbergiana Biol.* 40(5/6): 112–147.

MERTENS, R. 1959. Liste der Warane Asiens und der Indo-australischen Inselwelt mit systematischen Bemerkungen. *Senckenbergiana Biol.* 40(5/6): 221–240; 5 pl.

MERTENS, R. 1961. *Lanthonotus*: An important lizard in evolution. *Sarawak Mus. J.* 10: 283–285.

MERTENS, R. 1961. Was ist "*Naja celebensis* Ahl" in Wirklichheit? *Senckenbergiana Biol.* 42: 419–420.

MERTENS, R. 1963. Liste der rezenten Amphibien und Reptilien. Helodermatidae, Varanidae, Lanthonotidae. Das Tierreich 79. Walter de Gruyter & Co., Berlin. ix + 26 pp.

MERTENS, R. 1964. Weitere Mitteilungen über die Rassen von *Ablepharis boutonii* (Desjardin). III. *Zool. Anz.* 173: 99–110.

MERTENS, R. 1965. Eine neue Art der Fröschgattung *Oreophryne* von der Insel Rintja. *Senckenbergiana Biol.* 46(3): 189–191.

MERTENS, R. 1965. *Elaphe* Fitzinger 1833, nicht *Elaphis* Bonaparte 1831 (Reptilia, Serpentes). *Senckenbergiana Biol.* 46(3): 193–194.

MERTENS, R. 1966. The keeping of Borneo earless monitor (*Lanthonotus borneensis*). *Sarawak Mus. J.* 14: 320–322.

MERTENS, R. 1967. Die herpetologische Sektion des Natur-Museums und Forschungs-Institutes Senckenberg in Frankfurt a. M. nebst einem Verzeichnis ihrer Typen. *Senckenbergiana Biol.* 48(A): 1–106.

MERTENS, R. 1968. Die Arten und Unterarten der Schmuckbaumschlangen (*Chrysopelea*). *Senckenbergiana Biol.* 49(3/4): 191–217.

MERTENS, R. 1969. Über die Variabilitat der Achtstreifennater, *Oligodon octolineatus*. *Senckenbergiana Biol.* 50(5/6): 339–345.

MERTENS, R. 1970. Zur Frage der "Fluganpassungen" von *Chrysopelea* (Serpentes, Colubridae). *Salamandra* 6: 11–14.

MERTENS, R. 1971. Die Stachelschildkröte (*Heosemys spinosa*) und ihre Verwandten. *Salamandra* 7(2): 49–54.

MERTENS, R. 1971. Zur Kenntnis des javanischen Baumskinks, *Dasia leucosticta*. *Senckenbergiana Biol.* 52(3/5): 192–196.

MERTENS, R. & H. WERMUTH. 1955. Die rezenten Schildkröten, Krokodile und Brüchenechsen. *Zool. Jb. Abt. Allg. Syst.* 83(5): 323–440.

MEYERS, G. S. 1943. The lizard names *Platyurus* and *Cosymbotus*. *Copeia* 1943(3): 192.

MEYLAN, P. A. 1987. The phylogenetic relationships of soft-shelled turtles (Family Trionychidae). *Bull. American Mus. nat. Hist.* 186(1): 1–101.

MICHAUX, B. 1995. Distributional patterns in West Wallacea and their relationship to regional tectonic structures. *Sarawak Mus. J.* 48: 163–179.

MILLER, M. J. 1979. Ovophagia in the mourning gecko, *Lepidodactylus lugubris*. *Bull. Chicago Herpetol. Soc.* 14(4): 117–118.

MINTON, S. A., Jr. 1978. Observations on the Palawan mangrove snake, *Boiga dendrophila multicincta* (Reptilia, Serpentes, Colubridae). *J. Herpetol.* 12(1): 105–107.

MINTON, S. A., Jr. 1975. Geographic distribution of sea snakes. *In*: The biology of sea snakes. pp: 21–31. W. A. Dunson (Ed.). University Park Press, Baltimore.

MINTON, S. A., Jr. 1990. Venomous bites by nonvenomous snakes: an annotated bibliography of colubrid envenomation. *J. Wilderness Med.* 1(2): 119–127.

MINTON, S. A., Jr. & M. S. DA COSTA. 1975. Serological relationships of sea snakes and their evolutionary implications. *In:* The biology of sea snakes. pp: 33–55. W. A. Dunson (Ed.). University Park Press, Baltimore.

MINTON, S. A., Jr., H. G. DOWLING & F. E. RUSSELL. 1966. Poisonous snakes of the world. A manual for use by U.S. amphibious forces. Department of the Navy, Bureau of Medicine and Surgery, NAV-MED P-5099. Government Printing Office, Washington, D.C. viii + 212 pp; 81 pl.

MITCHELL, F. J. 1965. Australian geckos assigned to the genus *Gehyra* Gray. *Senckenbergiana Biol.* 46(4): 287–319.

MITTERMEIER, R. A. 1980. Conservation in Brunei. *Brunei Mus. J.* 4(4): 251–261.

MITTLEMAN, M. B. 1952. A generic synopsis of the lizards of the subfamily Lygosominae. *Smithsonian Misc. Coll.* 117(17): 1–35.

MOCQUARD, F. 1890. Récherches sur la faune herpetologique des Iles de Bornéo et de Palawan. *Nouv. Arch. Mus. Nat. Hist. Nat. Ser. 3* 2: 115–168.

MOCQUARD, F. 1892. Voyage de M. Chaper à Bornéo. Nouvelle contribution à la faune herpetologique de Bornéo. *Mem. Soc. Zool. France* 5: 140–206, 7 pl.

MOCQUARD, F. 1892. Nouvelle contribution à la faune herpétologique de Bornéo. *Mem. Soc. Zool. France* 5: 190–206.

MODIGLIANI, E. 1889. Materiali per la fauna erpetologica dell isola Nias. *Ann. Mus. Civ. Stor. nat. Genova* 7: 113–124; 1 pl.

MOEHN, L. D. 1984. Courtship and copulation in the Timor monitor, *Varanus timorensis. Herpetol. Rev.* 15(1): 14–16.

MOGK, C. W. F. 1860. Reptiliën van Banka. *Natuur. Tijd. Ned-Indië* 22(2): 97.

MOHANRAJ, P. & K. VEENAKUMARI. 1996. Perspectives on the zoogeography of the Andaman and Nicobar Islands, India. *Malayan Nat. J.* 50: 99–106.

MOLL, E. O. 1989. *Indotestudo forstenii* Travancore tortoise. *In:* The conservation biology of tortoises. pp: 118. I. R. Swingland & M. W. Klemens (Eds.). Occ. Pap. IUCN Species Survival Commission No. 5. IUCN, Gland.

MOLL, E. O. 1989. *Manouria emys* Asian brown tortoise. *In:* The conservation biology of tortoises. pp: 119–120. I. R. Swingland & M. W. Klemens (Eds.). Occ. Pap. IUCN Species Survival Commission No. 5. IUCN, Gland.

MONK, A. R. 1991. A case of mild envenomation from a mangrove snake bite. *Litteratura Serpentium* 11(1): 21–23.

MOODY, S. M. 1980. Phylogenetic and historical biogeographical relationships of the genera in the family Agamidae (Reptilia: Lacertilia). Ph.D. Dissertation, University of Michigan, Ann Arbor. 5 unnumbered pages + xv + 373 pp.

MORGAN, E. C. 1973. Snakes of the subfamily Sibynophiinae. Unpublished Ph.D. Dissertation, University of Southwestern Louisiana, Lafayette, Louisiana. 260 pg.

MORI, A., K. ARAYA & T. HIKIDA. 1995. Biology of the poorly known Bornean lizard, *Apterygodon vittatus* (Squamata: Scincidae): an arboreal anteater. *Herpetol. nat. Hist.* 3(1): 1–14.

MORI, A. & T. HIKIDA. 1991. Notes on the defensive behavior of the Asian elapid, *Maticora intestinalis. The Snake* 23(2): 107–109.

MORI, A. & T. HIKIDA. 1992. A preliminary study of sexual dimorphism in wing morphology of five species of the flying lizards, genus *Draco. Japanese J. Herpetol.* 14(4): 178–183.

MORI, A. & T. HIKIDA. 1993. Natural history observations of the flying lizard, *Draco volans sumatranus* (Agamidae, Squamata) from Sarawak, Malaysia. *Raffles Bull. Zool.* 41(1): 83–94.

MORI, A. & T. HIKIDA. 1994. Field observations on the social behavior of the flying lizard, *Draco volans sumatranus*, in Borneo. *Copeia* 1994(1): 124–130.

MORIGUCHI, H. 1988. A case of food item of a seasnake, *Laticauda laticaudata. The Snake* 20: 163.

MOULTON, J. C. 1922. The reported occurrence of Russell's viper in Sumatra & the Malay Peninsula. *J. Str. Br. Asiatic Soc.* 85: 206–207.

MUDDE, P. 1981. De Amboinese doorsschildpad (*Cuora amboinensis*). *Lacerta* 39(6 & &): 65–67.

MUDDE, P. 1982. *Cuora amboinensis* de Amboinese doorsschildpad. *Lacerta* 40(10 & 11): 261–263.

MÜLLER, F. 1895. Reptilien und Amphibien aus Celebes (II. Bericht). *Verh. Naturf. Ges. Basel* 10(3): 825–869.

MÜLLER, H. W. 1991. *Gonyosoma oxycephalum* (Boie). *Sauria, Berlin* Suppl. 13(1/4): 197–200.

MÜLLER, L. 1906. *Geoemyda spinosa* Gray. *Wschr. Aquar.-Terrar-Kde., Braunschweig* 3: 195–197; 207–209.

MÜLLER, S. & H. SCHLEGEL. 1845. Over de, in den Indischen Archipel voorkomende soorten van het slangengeslacht *Trigonocephalus*. pp: 49–58; pl. 7. *In:* C. J. Temminck. Verhandelingen over de natuurlijke geschiedenis der Nederlandsche overzeesche bezittingenis, door de leden der Natuurkundige Commissie in Indië en andere Schrijvers. S. & J. Luchtmans & C. C. Van de Hoek, Leiden.

MÜLLER, S. & H. SCHLEGEL. 1845. Over de Brilslangen van Indischen Archipel. pp: 69–72; pl. 10. *In:* C. J. Temminck. Verhandelingen over de natuurlijke geschiedenis der Nederlandsche overzeesche bezittingenis, door de leden der Natuurkundige Commissie in Indië en andere Schrijvers. S. & J. Luchtmans & C. C. Van de Hoek, Leiden.

MÜLLER, S. & H. SCHLEGEL. 1845. Over de Krokodillen van den Indischen Archipel. 28 pp., pl. 1–3. *In:* C. J. Temminck. Verhandelingen over de natuurlijke geschiedenis der Nederlandsche overzeesche bezittingenis, door de leden der Natuurkundige Commissie in Indië en andere Schrijvers. S. & J. Luchtmans & C. C. Van de Hoek, Leiden.

MURPHY, J. B. 1977. An unusual method of immobilizing avian prey by the dog-tooth cat snake. *Copeia* 1977(1): 182–184.

MURPHY, J. C. 1988. An overview of the Asian file snakes, Family Acrochordidae. *Bull. Chicago Herpetol. Soc.* 23(1): 1–4.

MURPHY, J. C. 1992. Fifty days of amphibian and reptile collecting in Sabah: A personal adventure with the fauna of Borneo. *Bull. Chicago Herpetol. Soc.* 27(2): 25–37.

MURPHY, J. C. & H. K. VORIS. 1994. A key to the homolopsine snakes. *The Snake* 26(2): 123–133.

MURPHY, J. C. & H. K. VORIS. 1994. Neglected serpents of Asian wetlands. *Asian Wetland News* 7: 20–22.

MURPHY, J. C., H. K. VORIS & D. R. KARNS. 1993. A key to the snakes of Sabah's Danum Valley. *Sabah Soc. J.* 10: 57–76.

MURPHY, J. C., H. K. VORIS & D. R. KARNS. 1994. A field guide and key to the snakes of the Danum Valley, a Bornean tropical forest ecosystem. *Bull. Chicago Herpetol. Soc.* 29(7): 133–151.

MURTHY, T. S. N. 1983. Rediscovery of the blind snake *Typhlops oatesii* in Andamans, India. *The Snake* 15: 48–49.

MURTHY, T. S. N. 1987. Curious reptiles of Andamans. *Sci. Reporter* 18: 32–36.

MUSTERS, C. J. M. 1983. Taxonomy of the genus *Draco* L. (Agamidae, Lacertilia, Reptilia). *Zool. Verh.* (199): 1–120.

NAINGGOLAN, F. J. 1929. Over de giftigheid van eenige slangen voor grootere zoogdieren, inzonderheid den mensch. *Trop. Natuur* 18: 189–192.

NAINGGOLAN, F. J. 1930. Iets over de „Oelar mati-ikoer" van de Atjeher af de „Tjintamanis" van de Minangkabauer. *Trop. Natuur* 19: 138–141.

NAINGGOLAN, F. J. 1933. Is het gif van de „Oelar welang" (*Bungarus candidus*) doodelijk voor andere slangen. *Trop. Natuur* 22: 134–136.

NAINGGOLAN, F. J. 1937. Iets over *Trimeresurus wagleri*. *Trop. Natuur* 26(11): 177–180.

NETSCHER, E., E. F. MEIJER & H. RAAT. 1859. Reptilien en visschen van Bintang. *Natuur. Tijd. Ned-Indië* 16(2): 45–46.

NEUHAUS, H. 1935. Neunachweis von *Vipera russellii* auf Java. *Treubia* 15(1): 49–50.

NG, P. 1991. Frogs of Sabah [Book Review]. *Raffles Bull. Zool.* 39(2): 375–376.

NIEDEN, F. 1923. Anura I. Das Tierreich 46. Walter de Gruyter & Co. X–XII + 1–584 pp.

NIEDEN, F. 1926. Anura II. Das Tierreich 49. Walter de Gruyter & Co. X–VI + 1–110 pp.

NODZENSKI, E., R. WASSERSUG & R. F. INGER. 1989. Developmental differences in visceral morphology of megophryne pelobatid tadpoles in relation to their body form and mode of life. *Biol. J. Linn. Soc.* 38: 369–388.

NUITJA, I. N. S. 1981. Incubation and hatching rate in the turtle *Chelonia mydas* Linnaeus. *J. Mar. Biol. Assoc. India* 23(1 & 2): 29–35.

NUITJA, I. N. S. 1982. Marine turtle nesting in Indonesia. *Copeia* 1982(3): 708–710.

NUITJA, I. N. S. 1996. Status of freshwater turtles in Java Island. *In*: International Congress of Chelonian Conservation. Proceedings. pp: 66. SOPTOM (Ed.). Editions Soptom, Gonfaron.

NUITJA, I. N. S. & S. AKMAD. 1982. Management and conservation of marine turtles in Indonesia. Paper presented at the Third World National Parks Congress, Bali. 16 pg.

NUITJA, I. N. S. & I. UCHIDA. 1983. Studies in the sea turtles-II. The nesting site characteristics of the hawksbill and the green turtles. *Treubia* 29(1): 63–79.

OBST, F. J. 1983. Zur Kenntnis der Schlangengattung *Vipera* (Reptilia: Serpentes: Viperidae). *Zool. Abh. Staatliches Mus. Tierkunde, Dresden* 38(13): 229–235.

OBST, F. J. 1983. Beiträg zur Kenntnis der landschildkröten-Gattung *Manouria* Gray, 1852 (Reptilia, Testudines, Testudinidae). *Zool. Abh. Staatliches Mus. Tierkunde, Dresden* 38(15): 247–256.

OESMAN, H. 1967. Breeding behavior of *Varanus komodoensis* at the Jogjakarta Zoo. *Int. Zoo Yrbk* 7: 10–11.

O'SHEA, M. 1985. The Borneo cave gecko (*Cyrtodactylus cavernicolus* Inger & King 1961)—its rediscovery on the Niah Caves of Sarawak. *The Herpetile* 10(3): 68–72.

OTA, H. & T. HIKIDA. 1988. A new species of *Lepidodactylus* (Sauria: Gekkonidae) from Sabah, Malaysia. *Copeia* 1988(3): 616–621.

OTA, H. & T. HIKIDA. 1988. Karyotypes of two species of the genus *Ptychozoon* (Gekkonidae: Lacertilia) from southeast Asia. *Japanese J. Herpetol.* 12(4): 139–141.

OTA, H. & T. HIKIDA. 1989. Karyotypes of three species of the genus *Draco* (Agamidae: Lacertilia) from Sabah, Malaysia. *Japanese J. Herpetol.* 13(1): 1–6.

OTA, H. & T. HIKIDA. 1991. Taxonomic review of the lizards of the genus *Calotes* Cuvier 1817 (Agamidae: Squamata) from Sabah, Malaysia. *Trop. Zool.* 4: 179–192.

OTA, H. & T. HIKIDA. 1996. The second specimen of *Calotes kinabaluensis* de Grijs (Squamata: Agamidae) from Sabah, Malaysia, with comments on the taxonomic status of the species. *J. Herpetol.* 30(2): 288–291.

OTA, H., T. HIKIDA & M. HASEGAWA. 1995. Karyotypes of two lygosomine lizards of the genus *Emoia* (Squamata: Scincidae) from Malaysia and Micronesia. *Russian J. Herpetol.* 2(1): 43–45.

OTA, H., T. HIKIDA, M. HASEGAWA, D. LABANG & J. NABHITABHATA. 1996. Chromosomal variation in the scincid genus *Mabuya* and its arboreal relatives (Reptilia: Squamata). *Genetica* 98: 87–94.

OTA, H., T. HIKIDA, M. KON & T. HIDAKA. 1989. Unusual nest site of a scincid lizard *Sphenomorphus kinabaluensis* from Sabah, Malaysia. *Herpetol. Rev.* 20(2): 38–39.

OTA, H., T. HIKIDA & M. MATSUI. 1991. Re-evaluation of the status of

Gecko verreauxi Tytler, 1864, from the Andaman Islands, India. *J. Herpetol.* 25(2): 147-151.

OTA, H., T. HIKIDA, M. MATSUI & A. MORI. 1990. Karyotype of *Gekko monarchus* (Squamata: Gekkonidae) from Sarawak, Malaysia. *Japanese J. Herpetol.* 13(4): 136-138.

OTA, H., T. HIKIDA, M. MATSUI & A. MORI. 1991. Karyotypes of two water skinks of the genus *Tropidophorus* (Reptilia: Squamata) from Borneo. *J. Herpetol.* 25(4): 488-490.

OTA, H., T. HIKIDA, M. MATSUI & A. MORI. 1992. Karyotypes of two species of the genus *Cyrtodactylus* (Squamata: Gekkonidae) from Sarawak, Malaysia. *Caryologia* 45(1): 43-49.

OTA, H., M. MATSUI, T. HIKIDA & T. HIDAKA. 1987. Karyotype of a gekkonid lizard, *Cosymbotus platyurus*, from Sabah, Borneo, Malaysia. *Zool. Sci.* 4: 385-386.

OTA, H., M. MATSUI, T. HIKIDA & A. MORI. 1992. Extreme karyotypic divergence between species of the genus *Gonocephaus* (Reptilia: Squamata: Agamidae) from Borneo and Australia. *Herpetologica* 48(1): 1201-124.

OTA, H., S. SENGOKU & T. HIKIDA. 1996. Two new species of *Luperosaurus* (Reptilia: Gekkonidae) from Borneo. *Copeia* 1996(2): 433-439.

OUWENS, P. A. 1912. On a large *Varanus* species from the island of Komodo. *Bull. Jard. Bot. Buit.* 2(6): 1-3.

PADIAN, K. & P. E. OLSEN. 1984. Footprints of the Komodo monitor and the trackways of fossil reptiles. *Copeia* 1984(3): 662-671.

PANDE, P., A. KOTHARI & S. SINGH (Eds.). 1991. Directory of National Parks and Sanctuaries in Andaman and Nicobar Islands. Management status and profiles. Indian Institute of Public Administration, New Delhi. 171 pp.

PARKER, H. W. 1924. Description of a new agamid lizard from Sumatra. *Ann. & Mag. nat. Hist. Ser. 9* 24: 624-625.

PARKER, H. W. 1926. The brevicipitid frogs of the genus *Microhyla*. *Ann. & Mag. nat. Hist. Ser. 10* 2: 473-499.

PARKER, H. W. 1934. A monograph of the frogs of the family Microhylidae. British Museum (Natural History), London. vi + 208 pp.

PASTEUR, J. D. 1941. Iets over Slangen van Zuid-Sumatra. *Trop. Natuur* 30: 85-88.

PEEL, G. 1981. *Boiga dendrophila melanota* de mangrove nachtboomslang. *Litteratura Serpentium* 1(4): 132-138.

PERACCA, M. G. "1899" 1900. Reptiles et batrachiens de l'Archipel Malais. *Rev. Suisse Zool.* 7: 321-330.

PERÄLÄ, J. 1996. Herpetologisia havaintoja Malesian niemimaalta, Sarawakista, Sabahista ja Singaporesta. Osa II: Penang. *Herpetomania* 5(2): 26-29.

PERÄLÄ, J. 1996. Herpetologisia havaintoja Malesian niemimaalta, Sarawakista, Sabahista ja Singaporesta. Osa III: houmioita merinahkakakilpikonnan soujelusta rantau abangissa (Terengganu) [Herpetological field studies in Singapore, Peninsular Malaysia, Sabah and Sarawak 1994–95. Part III: *Dermochelys coriacea* in Rantau Abang (Terengganu)]. *Herpetomania* 5(3–4): 51–57.

PETERS, U. W. 1989. Kurz vorgestellt: Drei Wasserschildkröten. *Platysternon megacephalum, Cuora trifasciata* und *Notochelys platynota. Das Aquarium* 23(246): 777–778.

PETERS, W. C. H. 1859. Nachrichten von Hrn. Fedor Jagor, der auf eigne Kosten nach Ostindien und des Philippinen gereist und von der Akademie mit Instructionen für das Sammeln naturwissenschaffliches Gegenstände versehen wurde. *Mber. Königl. Akad. Wiss. Berlin* 1859: 269–271.

PETERS, W. C. H. 1860. Über einige interessante Amphibien, welche von dem durch seine zoologischen Schriften rühmlichst bekannten österreichischen Naturforscha Professor Schmarda während seiner auf mehrere Welttheile ausgedehnten, besonders auf wirbellose Thiere gerichteten, naturwissenschafflicken Reise, mit deren Veröffentlichung Hr. Schmarda gegenwärtig in Berlin beschaffgt ist, auf der Insel Ceylon gesammelt wurden. *Mber. Königl. Akad. Wiss. Berlin* 1860: 182–186.

PETERS, W. C. H. 1861. Eine zweite Übersicht (vergl. Monatsberichte 1859 p. 269) der von Hrn. F. Jagor auf Malacca, Java, Borneo und den Philippinen gesammelten und dem Kgl. zoologischen Museum übersanden Schlangen. *Mber. Königl. Akad. Wiss. Berlin* 1861: 683–691.

PETERS, W. C. H. 1862. Präparate von zur craniologischen Unterscheidung der Schlangengattung *Elaps* und über eine neue Art der Gattung *Simotes, S. semicinctus. Mber. Königl. Akad. Wiss. Berlin* 1862: 635–638.

PETERS, W. C. H. 1864. Eine junge *Caecilia glutinosa* (*Epicrinium hypocyaneum*) mit Kemenlöchern aus Malacca. *Mber. Königl. Akad. Wiss. Berlin* 1864: 303–304.

PETERS, W. C. H. 1864. Über einige Säugethiere (*Mormops, Macrotus, Vesperus, Molossus, Capromys*), Amphibien (*Platydactylus, Otocryptis, Euprepes, Ungalia, Dromicus, Tropidonotus, Xenodon, Hylodes*) und Fische (*Sillago, Sebastes, Channa, Myctophum, Carassius, Barbus, Capoëta, Poecilia, Saurenchelys, Leptocephalus*). *Mber. Königl. Preuss. Akad. Wissenschaft. Berlin* 1864: 381–399.

PETERS, W. C. H. 1867. Herpetologische Notizen. *Mber. Königl. Akad. Wiss. Berlin* 1867: 13–37.

PETERS, W. C. H. 1871. Über neue Reptilien aus Ostafrica und Sarawak (Borneo), vorzüglich aus der Sammlung des Hern. Marquis J. Doria zu Genua. *Mber. Königl. Akad. Wiss. Berlin* 1871: 566–581.

PETERS, W. C. H. 1872. Übersicht der von den Herren M. G. Doria und D. O. Beccari in Sarawak auf Borneo von 1865 bis 1868 gesammelten Amphibien. *Ann. Mus. Civ. Stor. Nat. Genoa Ser. 1* 3: 27–45.

PETERS, W. C. H. 1872. Über einige von Hrn. Dr. A. B. Meyer bej Gorontalo und auf den Togian-Inseln gesammelte Amphibien. *Mber. Königl. Akad. Wiss. Berlin* 1872: 581–585.

PETERS, W. C. H. & G. DORIA. 1878. Catalogo dei rettili e dei batraci raccolti da O. Beccari, L. M. D'Albertis e A. A. Bruijn nella sotto-regione Austro-Malese. *Ann. Mus. Civ. Stor. Nat. Genova Ser. 1* 13: 323–450.

PETZOLD, H.-G. 1974. Erfolg und Misserfolg mit Javanischen Flugdrachen, *Draco volans. Aquar. Terr.* 21: 158–163.

PFEFFER, P. 1959. Observations sur le varan de Komodo (*Varanus komodoensis* Ouwens, 1912). *Terre Vie* 106(2/3): 195–243.

PFEFFER, P. 1962. Parades et compartement territorial chez trois espèces de Dragons-volants (Agamides) d'Indonésie. *Terre Vie* 109: 417–427.

PFEFFER, P. 1965. Aux Iles du dragon. Flammarion, Paris. 203 pp.

PHILIPPEN, H.-D. 1994. Das Porttrait: *Liasis mackloti savuensis* Brongersma, 1956. *Sauria, Berlin* 16(1): 2.

PHILLIPPS, S. 1981–82. An uninvited guest. *Sabah Soc. J.* 7(2): 136–137.

PIETERS D. 1933. Een geluidmakende slang (*Acrochordus javanicus* Hornst.). *Trop. Natuur* 22: 13–15.

PILLAI, R. S. 1977. On two frogs of the family Microhylidae from Andamans including a new species. *Proc. Indian Acad. Sci.* 86: 135–138.

PILLAI, R. S. 1991. Contribution to the amphibian fauna of Andaman and Nicobar with a new record of the mangrove frog, *Rana cancrivora. Rec. Zool. Surv. India* 88(1): 41–44.

POLUNIN, N. V. C. 1975. Sea turtle reports on Thailand, West Malaysia, and Indonesia with a synopsis of data on the conservation status in the Indo-West Pacific region. Report to the IUCN, Morges. 40 pp.

POLUNIN, N. V. C. & I. N. S. NUITJA. 1982. Sea turtle populations of Indonesia and Thailand. *In*: Biology and conservation of sea turtles. pp: 353–362. K. A. Bjorndal (Ed.). Smithsonian Institution Press, Washington, D.C.

PREMO, D. B. 1985. The reproductive ecology of a ranid frog community in pond habitats of West Java, Indonesia. Unpublished Ph. D. dissertation, Michigan State University.

PREMO, D. B. & A. H. ATMOWIDJOJO. 1987. Dietary patterns of the "crab-eating frog", *Rana cancrivora*, in west Java. *Herpetologica* 43(1): 1–6.

PRITCHARD, P. C. H. 1979. Encyclopedia of turtles. T. F. H. Publications, Inc. Ltd., Neptune, New Jersey. 895 pp.

PRITCHARD, P. C. H. 1980. *Dermochelys, D. coriacea. Cat. American Amphibians & Reptiles* 238: 1–4.

PRITCHARD, P. C. H. 1993. Carapacial pankinesis in the Malayan softshell turtle, *Dogania subplana. Chelonian Conserv. & Biol.* 1(1): 31–36.

PROCTER, J. B. 1929. On a living Komodo dragon (*Varanus komodoensis* Ouwens) exhibited at the scientific meeting, October 23, 1928. *Proc. Zool. Soc. London* 23: 1017–1019.

PROUD, K. R. S. 1978. Some notes on a captive earless monitor lizard, *Lanthonotus borneensis. Sarawak Mus. J.* 26: 235–242.

QUINN, H. R. & K. NEITMAN. 1976. Reproduction in the snake *Boiga cynodon* (Reptilia, Serpentes, Colubridae). *J. Herpetol.* 12(2): 255–256.

RABOR, D. S. 1981. Philippine reptiles and amphibians. Pundasyon sa

Pagpapaunlad ng Kaalaman sa Pagtu turo ng Agham, Ink, Quezon City. vi + 79 pp.

RADFORD, L. & F. L. PAINE. 1989. The reproduction and management of the Dumeril's monitor, *Varanus dumerili*, at the Buffalo Zoo. *Int. Zoo Yrbk* 28: 153-155.

RANJITSINH, M. K. 1983. Ora: the Komodo dragon—a short natural history. *Sanctuary Asia* 3(2): 132-139.

RAO, D.-Q. & D. T. YANG. 1992. [Phylogenetic systematics of Pareinae (Serpentes) of southeast Asia and adjacent islands with relationship between it and the geology changes.] *Acta Zool. Sinica* 38(2): 139-150. [In Chinese with English abstract.]

RAO, G. C. 1987. Sea turtles in the Bay Islands. *The Daily Telegrams* 189: 8-17.

RAO, G. C. & M. K. DEV ROY. 1985. The fauna of the Bay Islands. *J. Andaman Sci. Assoc.* 1(1): 1-17.

RAO, G. C. & I. H. KHAN. 1989. On the present status of the marine fauna of the Andaman Sea. *Zoologiana* 5: 29-42.

RAO, G. C. & H. S. MEHTA. 1986. Frogs and toads in Bay Islands. *The Daily Telegrams* 186: 12-19.

RAPPARD, F. W. 1936. Een schildpadenstrand in het wild reservaat zuid Sumatra. *Trop. Natuur* (Jubileum Uitgave): 124-126.

RASMUSSEN, J. B. "1997" 1996. Systematics of sea snakes: a critical review. *In*: Venomous snakes: ecology, evolution and snakebite. pp: 15-30. R. S. Thorpe, W. Wüster & A. Malhotra (Eds.). Symp. Zool. Soc. London (70). The Zoological Society of London/Clarendon Press, Oxford.

RASMUSSEN, J. B. 1975. Geographical variation, including an evolutionary trend in *Psammodynastes pulverulentus* (Boie, 1827)(Boiginae, Homalopsidae, Serpentes). *Vidensk. Meddr. Dansk. Naturh. Foren.* 138: 39-64.

RASMUSSEN, J. B. 1979. An intergeneric analysis of some boigine snakes —Bogert's (1940) Group XIII & XIV (Boiginae, Serpentes). *Vidensk. Meddr dansk naturh. Foren.* 141: 97-155.

RASMUSSEN, J. B. 1990. The retina of *Psammodynastes pulverulentus* (Boie, 1827) and *Telescopus fallax* (Fleischmann, 1831) with a discussion of their phylogenetic significance (Colubroidea, Serpentes). *Z. Zool. Syst. Evolut.-forsch.* 28: 269-276.

RASMUSSEN, J. B. & A. F. STIMSON. 1983. *Boiga* Fitzinger, 1826 (Reptilia, Serpentes), proposed conservation under the plenary powers. Z. N.(S.)2404. *Bull. Zool. Nomen.* 40(4): 209-210.

RATNAM, J. 1992. Distribution and behavioural ecology of the Andaman day gecko (*Phelsuma andamanensis*). Masters Dissertation, Sálim Ali School of Ecology, Pondicherry University, Pondicherry. 110 pg.

RATNAM, J. 1993. Status and natural history of the Andaman day gecko, *Phelsuma andamanensis. Dactylus* 2(2): 59-66.

REGENASS, U. & E. KRAMER. 1981. Zur Systematik der grüner Grubenottern der Gattung *Trimeresurus* (Serpentes, Crotalidae). *Rev. Suisse Zool.* 88(1): 163–205.

REHÁK, I. 1990. [Reproductive biology of *Gonyosoma oxycephalum* in captivity.] *Akvárium Terárium, Praha* 33(3): 30–32. [In Polish.]

REID, J. A. 1958. Gliding by the green crested lizard. *Malayan nat. J.* 12: 119.

REIDE, K. 1995. Räumliche und zeitliche Strukturierung tropischer Rufgemeinschaften. *In*: Verhandlungen der deutschen Zoologischen Gesselschaft 88.1. pp: 45. Gustav Fischer Verlag, Stuttgart.

REIDE, K. 1996. Diversity of sound-producing insects in a Bornean lowland rain forest. *In:* Tropical Rainforest Research: Current Issues. pp: 77–84. D. S. Edwards, W. E. Booth & S. C. Choy (Eds.). Kluwer Academic Publishers, Dordrecht.

REINHARDT, J. C. H. 1836. Afhandling om *Xenodermus javanicus. K. dansk. Vidensk. Selk. Noturvid. Math. Afh. Kjobenhavn* 3: 6–7.

REINHARDT, J. C. H. 1843. Beskrivelse af nogle nye Slangearter. *K. dansk. Vidensk. Selk. Noturvid. Math. Afh. Kjobenhavn* 10: 233–279; pl. 1–3.

REKOSWARDOJO, R. 1961. Penyu dipantai pengumbahan. *Penggemar Alam* 40(1): 16–24.

RESE, R. 1986. *Varanus dumerilii* (Schlegel). *Sauria, Berlin* 8(1): 2.

RESE, R. 1986. Der Kommentkampf bei *Varanus salvator. Sauria, Berlin* 8(2): 27–29.

RHODIN, A. G. J., P. C. H. PRITCHARD & R. A. MITTERMEIER. 1984. The incidence of spinal deformities in marine turtles, with notes on the prevalence of kyphosis in Indonesian *Chelonia mydas. British J. Herpetol.* 6: 369–373.

RHODIN, A. G. J. & H. M. SMITH. 1982. The original authorship and type specimen of *Dermochelys coriacea. J. Herpetol.* 16(3): 316–317.

RIQUIER, M. A. 1997. Population ecology of the water monitor (*Varanus salvator*) in West Kalimantan, Indonesia. *In*: Herpetology '97. Abstracts of the Third World Congress of Herpetology. 2–10 August, 1997. pp: 173. Z: Rocek & S. Hart (Eds). Third World Congress of Herpetology, Prague.

ROBINSON, H. C. & C. B. KLOSS. 1918–1923. A nominal list of the species of reptiles and batrachians occurring in Sumatra. Addenda and corrigenda. *J. Fed. Malay States Mus.* 8, June 1918–November 1923: 297–308; 362–364 and 365.

ROGNER, M. 1995. Schildkröten 1. Chelydridae Dermatemydidae Emydidae. Heidi Rogner-Verlag, Hürtgenwald. 192 pp.

ROGNER, M. 1996. Schildkröten 2. Kinosternidae Platysternidae Testudinidae Trionychidae Carettochelyidae Cheloniidae Dermochelyidae Chelidae Pelomedusidae. Heidi Rogner-Verlag, Hürtgenwald. 265 pp.

ROSS, C. A. 1990. *Crocodylus raninus* S. Müller and Schlegel, a valid species of crocodile (Reptilia: Crocodylia) from Borneo. *Proc. Biol. Soc. Washington* 103: 955–961.

ROSS, C. A. 1992. Designation of a lectotype for *Crocodylus raninus* S. Müller and Schlegel (Reptilia: Crocodylidae), the Bornean crocodile. *Proc. Biol. Soc. Washington* 105(2): 400–402.

ROSS, C. A., J. COX & H. KURNIATA. 1996. Crocodile studies in Kalimantan. *Crocodile Spec. Group Newsl.* 15(1): 7.

ROSSMAN, D. A. & W. C. EBERLE. 1977. Partition of the genus *Natrix*, with preliminary observations on evolutionary trends on natricine snakes. *Herpetologica* 33(1): 34–43.

ROSSMAN, N.J., D. A. ROSSMAN & N. K. KEITH. 1982. Comparative visceral topography of the New World snake tribe Thamnophiini (Colubridae, Natricinae). *Tulane Stud. Zool. Bot.* 23(2): 123–164.

ROUX, J. 1904. Reptilien und Amphibien aus Celebes. *Verh. Nat. Ges. Basel* 15(3): 425–433.

ROUX, J. 1911. Elbert-Sunda-Expedition des Frankfurter Vereins für Geographie und Statistik. Reptilien und Amphibien. *Zool. Jahrb.* 30(5): 495–508.

ROUX, J. 1914. Note sur une espèce nouvelle d'*Oligodon* provenant de Sumatra. *Rev. Suisse Zool.* 22(2): 27–29.

ROUX, J. 1925. Note sur une collection de reptiles et d'amphibiens de l'Ile Nias. *Rev. Suisse Zool.* 32: 319–321.

ROUX, J. 1935. Note sur deux gecko non encore signalés à Sumatra. *Verh. Naturf. Ges. Basel* 46: 56–58.

RUMMLER, H.-J. & FRITZ, U. 1991. Geographische variabilität der Amboina-Scharnierschildkröte *Cuora amboinensis* (Daudin, 1802), mit Beschreibung einer neuen unterart, *C. a. kamaroma* subsp. nov. *Salamandra* 27(1): 17–45.

RUTHVEN, A. G. 1921. *Oligodon rouxi*, a new name for *Oligodon ornatus* Roux. *Copeia* 1921(1): 20.

SACHS, W. B. 1927. Neues und Zusammengefassten vom Riesenwaren der Komodo Inseln, *Varanus komodoensis* Ouwens. *Blat. Aquar.-Terr.* 38(22): 450–456.

SACKETT, J. T. 1940. Zoological results of the George Vanderbilt Sumatran Expedition, 1936–1939. Part 4: The reptiles. *Acad. Nat. Sci. Philadelphia* 41: 1–3.

SALVADOR, A. 1975. Un nuevo cecilido procedente de Java (Amphibia: Gymnophiona). *Bonn. Zool. Beitrg.* 26(4): 366–369.

SANCHEZ-HERRAIZ, M. J., R. MARQUEZ, L. J. BARBADILLO & J. BOSCH. 1995. Mating calls of three species of anurans from Borneo. *Herpetol. J.* 5(4): 293–297.

SARKAR, A. K. 1990. Taxonomic and ecological studies on the amphibians of Andaman and Nicobar Islands, India. *Rec. Zool. Surv. India* 86: 103–117.

SAVAGE, J. M. 1952. Two centuries of confusion: The history of the snake name *Ahaetulla*. *Bull. Chicago Acad. Sci.* 9: 203–216.

SAWYER, F. C. 1953. The dates of issue of J. E. Gray's "Illustrations of Indian Zoology" (London, 1830–1835). *J. Soc. Bibl. nat. Hist.* 3(1): 48–55.

SAXENA, A. 1994. Captive breeding of the Malayan box turtle (*Cuora amboinensis*) from the Nicobar Islands. *Hamadryad* 19: 93–94.

SCHAFER, C. 1992. *Gekko monarchus* (Duméril & Bibron). *Sauria, Berlin* Suppl. 14(1–4): 243–246.

SCHIJFSMA, K. 1932. Notes on some tadpoles, toads and frogs from Java. *Treubia* 14(1): 43–72.

SCHLEGEL, H. 1926. Notice sur l'Erpetologie de l'île de Java; par M. Boié (Ouvrage manuscript). *Bull. Sci. nat. Géol.* 9: 233–240.

SCHLEGEL, H. 1827. Erpetologische Nachzuchten. *Isis* 20(3): 281–294.

SCHLEGEL, H. 1837. Essai sur la physionomie des serpens. Vol. 1. Kips, Hz and van Stockum, Amsterdam and Arnz & Comp., Leide (= Leiden). xxviii + 251 pp.

SCHLEGEL, H. 1837. Essai sur la physionomie des serpens. Vol. 2. Kips, Hz and van Stockum, Amsterdam and Arnz & Comp., Leide (= Leiden). 606 + xiii pp.

SCHLEGEL, H. 1839. Abbildungen neuer oder unvollständig bekannter Amphibien, nach der Natur oder dem Leben entworfen, herausgegeben und mit einem erläuternden Texte begleitet. Arnz & Comp., Düsseldorf. xiv + 141 pp. (the dates of publication of the pages, as given by Williams and Wallach, 1989, are as follows: pp: 1–32 [1837]; 33–80 [1839] and 81–141 [1844].)

SCHLEGEL, H. & S. MÜLLER. 1839–1845. Over de schildpadden van Indischen Archipel en beschreijving eener nieuwe soort van Sumatra. *In*: Verhandlingen over de natuurlijke geschiedenis der Nederlandsche overzeesche bezittingen, door de Leden der Natuurkundige Commissie in Ost-Indië en andere Schrijvers. Zoologie. Vol. 3. pp: 29–36. C. J. Temminck (Ed.). Luchtmans & van der Hoek, Leiden.

SCHLEGEL, H. & S. MÜLLER. 1839–1845. Over de schlangensoorten van het geslacht *Homalopsis*, uit den Indischen Archipel. *In*: Verhandlingen over de natuurlijke geschiedenis der Nederlandsche overzeesche bezittingen, door de Leden der Natuurkundige Commissie in Ost-Indië en andere Schrijvers. Zoologie. Vol. 3. pp: 59–62. C. J. Temminck (Ed.). Luchtmans & van der Hoek, Leiden.

SCHLEGEL, H. & S. MÜLLER. 1839–1845. Over de slangen van het geslacht *Elaps*, welke den Indischen Archipel bewonen. *In*: Verhandlingen over de natuurlijke geschiedenis der Nederlandsche overzeesche bezittingen, door de Leden der Natuurkundige Commissie in Ost-Indië en andere Schrijvers. Zoologie. Vol. 3. pp: 59–62. C. J. Temminck (Ed.). Luchtmans & van der Hoek, Leiden.

SCHMIDT, A. 1976. Erst-Nachzucht des Zipfelfrösches *Megophrys nasuta*. *Salamandra* 12(2): 55–68.

SCHMIDT, A. 1978. Erst-nachzucht des Indischen Oschenfrösches *Kaloula pulchra*. *Salamandra* 14(2): 49–57.

SCHMIDT, A. 1994. Erst-Nachzucht der asiatischen Baumkröte *Pedostibes hosii*. *Salamandra* 30: 225–233.

SCHMIDT, A. & R. WICKER. 1977. Weitere Beobachtungen bei der Nachzucht des Zipfelfrösches *Megophrys nasuta* (Amphibia, Salientia, Pelobatidae). *Salamandra* 13(1): 43–48.

SCHMIDT, P. 1852. Beiträge zur ferneren Kenntniss der Meerschlangen. *Abh. Ges. Naturwiss., Hamburg* 2(2): 69–86.

SCHNEIDER, J. G. 1799. Historiae Amphibiorum naturalis et literariae. Fasciculcus Primus continens *Ranas, Calamitas, Bufones, Salamandras et Hydros* in genera et species descriptos notisque suis distinctos. Friederici Frommanni (sic for Fromann), Jena. xiii + (1) + 264 + (2) pp + pl. 1–2. Reprinted 1968 A. Asher.

SCHNEIDER, J. G.. 1801. Historiae Amphibiorum naturalis et literariae. Fasciculcus Secundus continens Crocodilos, Scincos, Chamaesauras, Boas, Pseudoboas, Elapes, Angues, Amphisbaenas et Caecilias. Friederici Fromman, Jena. vi + 374 pp + pl. 1–2. Reprinted 1968 A. Asher.

SCHOENMAKERS, J. 1935. Hoe de Komodo-varanen gevangen werden. *De Telegraaf (Surabaya)* 30 June: 3.

SCHOEPFF, J. D. 1792–1801. Historia Testudinum iconibus illustrata. Johannis Jacobi, Palm, Erlangae. xii + 136 pp. + Pl. I–XXII. Latin version of Naturgeschichte der Schildkroten mit Abbildungen erlautert. Johannis Jacobi Palm, Erlangen. (the dates of publication of the parts, acccording to Ernst et al., 1994 is: 1792: 1–32; 1793: 33–88; 1795: 89–136; 1801: 137–160.)

SCHULZ, K.-D. 1986. *Maticora bivirgata flaviceps* (Cantor, 1829). *Sauria, Berlin* 8(3): 2.

SCHULZ, K.-D. 1987. Die hinter-asiatischen Kletternattern der Gattung *Elaphe*. Teil 10: *Elaphe flavolineata* (Schlegel, 1837). *Sauria, Berlin* 9(3): 21–23.

SCHULZ, K.-D. 1987. *Elaphe taeniura* (Cope, 1861), a remarkable snake from Asia. *Litt. Serp.* 7(1): 4–16.

SCHULZ, K.-D. 1988. Die hinter-asiatischen Kletternattern der Gattung *Elaphe*. Teil 14. *Elaphe subradiata* (Schlegel, 1837) und *Elaphe erythrura* (Duméril & Bibron, 1854). *Sauria, Berlin* 10(3): 17–20.

SCHULZ, K.-D. 1989. Die hinter-asiatischen Kletternattern der Gattung *Elaphe*. Teil 17. *Elaphe porphyracea* (Cantor, 1839). *Sauria, Berlin* 11(4): 21–24.

SCHULZ, K.-D. 1996. A monograph of the colubrid snakes of the genus *Elaphe* Fitzinger. Koeltz Scientific Books, Havlickuv Brod, Czech Republic. iii + 439 pp.

SCHULZ, K.-D. & V. N. SCHEIDT. 1992. An introduction to the Indonesian rat snakes of the genus *Elaphe*. *Vivarium* 4(2): 13–17.

SCHULZ, K.-D. & H. SLEGERS. 1985. Erfolgreiche Haltung eines *Bungarus fasciatus* (Schneider, 1801). *Sauria, Berlin* 8(2): 3–4.

SCHWEIGGER, A. F. 1812. Prodromus monographiae cheloniorum. Königsberg. *Arch. Naturwiss. Math.* 1: 271–358, 406–458.

SCLATER, W. L. 1891. Notes on the collection of snakes in the Indian Museum with descriptions of several new species. *J. Bombay nat. Hist. Soc.* 60(3): 230–250; Pl. VI.

SCOTT, D. A. & C. M. POOLE. (Eds.). 1989. A status overview of Asian wetlands. Asian Wetland Bureau, Kuala Lumpur. 140 pp.

SCOTT, N.J. Jr. 1976. The abundance and diversity of the herpetofauna of tropical forests litter. *Biotropica* 8: 41–58.

SCRIVEN, K. 1972. Conservation in West Malaysia. *WWF Yrbk* 1971–72: 275–281.

SEAL, U.S., J. MANANSANG, D. SISWOMARTONO, T. SUHARTONO & J. SUGARJITO. 1997. Komodo monitor PHVA: Executive summary. *Australasian Herp News* (18): 5–7.

SEBA, A. 1734. Locupletissimi rerum naturalium thesauri accurata et descriptio, et iconibus artificiosissimus expressio, per universam physices historiam. Vol. I. Janssonio-Waesbergios, & J. Wetstenium, & Gul. Smith, Amsterdam. 178 pp + Pl. I–CXI.

SEBA, A. 1735. Locupletissimi rerum naturalium thesauri accurata et descriptio, et iconibus artificiosissimus expressio, per universam physices historiam. Vol. II. Janssonio-Waesbergios, & J. Wetstenium, & Gul. Smith, Amsterdam. 154 pp + Pl. I–CXIV.

SEBASTIAN, A. C. 1994. The Tomistoma, *Tomistoma schlegelii*, in southeast Asia, a status review and priorities for its conservation. *In*: Proceedings of the 12th Working Meeting of the Crocodile Specialist Group of the Species Survival Commission of IUCN-The World Conservation Union. 1: pp: 98–112. IUCN, Gland.

SEBASTIAN, A. C. 1994. Towards a priority listing for the conservation of crocodilians in the Indo-Malayan Realm. *Asian Wetland News* 7(1): 24–25.

SEKAR, A. G. 1984. Distribution of *Bufo camortensis* Mansukhani & Sarkar in the Andaman and Nicobar Islands. *J. Bombay nat. Hist. Soc.* 81(2): 488.

SEKI, T., S. KIKUYAMA & N. YANAIHARA. 1995. Morphology of the skin glands of the crab-eating frog (*Rana cancrivora*). *Zool. Sci.* 12(5): 623–626.

SHAW, G. 1802. General zoology, or systematic natural history. Vol. III. Part II. Amphibia. G. Kearsley, London. vii + 313–615; Pl. 87–140.

SHAW, G. & F. P. NODDER. 1797. Foeminae lectissimae et ornatissimae juliae ducissae northumbriensi nonum hunc naturae vivarii fasciculum. *Nat. Misc.* 8: [unpaginated]; Pl. 291.

SHAW, G. & F. P. NODDER. 1802. General zoology or systematic natural history. Vol. 3. Part. 2. Amphibia. G. Kearsley, London. v + pp 313–615 (= 303 pp.).

SHELFORD, R. W. C. 1901. A list of the reptiles of Borneo. *J. Straits Br. Royal Asiatic Soc.* (35): 43–68.

SHELFORD, R. W. C. 1901. On two new snakes from Borneo. *Ann. & Mag. nat. Hist. Ser. 7* 8: 516–517.

SHELFORD, R. W. C. 1902. A list of the reptiles of Borneo—Addenda et corrigenda. *J. Str. Br. Royal Asiatic Soc.* (38): 133–135.

SHELFORD, R. W. C. 1905. A new lizard and a new frog from Borneo. *Ann. & Mag. nat. Hist. Ser. 7* 15: 208–210.

SHELFORD, R. W. C. 1906. A note on "flying" snakes. *Proc. Zool. Soc. London* 1: 227–230.

SHELFORD, R. W. C. 1916. A naturalist in Borneo. T. Fisher Unwin Ltd., London. Reprinted, Oxford University Press, Singapore. 331 pp.

SHERBORN, C. D. 1895. On the dates on Shaw and Nodder's Naturalist's Miscellany. *Ann. & Mag. nat. Hist. Ser. 6* 15: 375–376.

SHETTY, S. 1996. Studies on the terrestrial behaviour of the yellow-lipped sea krait (*Laticauda colubrina*) in the Andaman Islands. M. Sc. Dissertation, Sálim Ali School of Ecology, Pondicherry University, Pondicherry. 44 pg + 2 maps.

SHETTY, S. & K. V. DEVI PRASAD. 1996. Studies on the terrestrial behaviour of *Laticauda colubrina* in the Andaman Islands, India. *Hamadryad* 21: 23–26.

SHETTY, S. & K. V. DEVI PRASAD. 1996. Geographic variation in the number of bands in *Laticauda colubrina*. *Hamadryad* 21: 44–45.

SHINE, R., P. HARLOW, J. S. KEOGH & (P.) BOEADI. 1995. Biology and commercial uilization of acrochordid snakes, with special reference to karung (*Acrochrodus javanicus*). *J. Herpetol.* 29(3): 352–360.

SHINE, R., P. HARLOW, J. S. KEOGH & (P.) BOEADI. 1996. Commercial harvesting of giant lizards. The biology of water monitor, *Varanus salvator* at South Sumatra. *Biol. Conserv.* 77: 125–134.

SHOEMAKER, V. H. & L. L. McCLANAHAN. 1980. Nitrogen excretion and water balance in amphibians of Borneo. *Copeia* 1980(3): 446–451.

SHUTTLEWORTH, C. 1981. Malaysia's green and timeless world. Heinemann Asia, Kuala Lumpur. 221 pp.

SIEDLECKI, M. 1909. Zur Kenntnis des javanischen Flugfrösches. *Biol. Zentralbl.* 29: 704–714; Pl. VII–VIII; 715–737; Pl. IX–X.

SIM, S. L. 1977. Snake bite and its management in Sarawak. *Sarawak Mus. J.* 25(46): 211–220.

SIVASUNDAR, A. 1995. Sea turtle survey of South Reef (June–July 1995). A preliminary report. Report to the Andaman and Nicobar Environmental Team/Madras Crocodile Bank Trust. 3 pg.

SIVASUNDAR, A. 1996. Studies on the nesting of leatherback sea turtles (*Dermochelys coriacea*) in the Andaman Islands. M. Sc. Dissertation, Sálim Ali School of Ecology, Pondicherry University, Pondicherry. 41 pg + 6 pl.

SIVASUNDAR, A. & K. V. DEVI PRASAD. 1996. Placement and predation of nest in leatherback sea turtles in the Andaman Islands, India. *Hamadryad* 21: 36–42.

SLOWINSKI, J. B. 1989. The interrelationships of laticaudine sea snakes based on the amino acid sequences of short-chain neurotoxins. *Copeia* 1989(3): 783–788.

SLOWINSKI, J. B. 1994. A phylogenetic analysis of *Bungarus* (Elapidae) based on morphological characters. *J. Herpetol.* 28(4): 440–446.

SMEDLEY, N. 1928. Some reptiles and Amphibia from the Anamba Islands. *J. Malayan Br. Royal Asiatic Soc.* 6(3): 76–77.

SMEDLEY, N. 1931. On some reptiles and a frog from the Natuna Islands. *Bull. Raffles Mus.* 5: 46–54.

SMEDLEY, N. 1931. Oviparity in a sea-snake, *Laticauda colubrina* (Schneid.). *Bull. Raffles Mus.* 5: 54–59.

SMEDLEY, N. 1931. Notes on the giant frog, *Rana macrodon. Bull. Raffles Mus.* 5: 59–62.

SMEDLEY, N. 1931. Amphibians and reptiles from the South Natuna Islands. *Bull. Raffles Mus.* 6: 102–104.

SMITH, H. M. 1969. The nomenclature of certain taxa of higher categories in snakes. *J. Herpetol.* 3(1–2): 19–25.

SMITH, H. M. 1989. The original description of *Ovophis* Burger (Serpentes: Viperidae). *Bull. Chicago Herpetol. Soc.* 24: 7.

SMITH, H. M & A. G. J. RHODIN. 1986. Authorship of the scientific name of the leatherback sea turtle. *J. Herpetol.* 20(3): 450–451.

SMITH, H. M., R. B. SMITH & H. L. SAWIN. 1977. A summary of snake classification (Reptilia, Serpentes). *J. Herpetol.* 11(2): 115–121.

SMITH, M. A. 1921. Two new batrachians and a new snake from Borneo and the Malay Peninsula. *J. Fed. Malay St. Mus.* 10: 197–199; Pl. II.

SMITH, M. A. 1925. On a collection of reptiles and amphibians from Mt. Murud, Borneo. *J. Sarawak Mus.* 3: 5–14.

SMITH, M. A. 1925. Contribution to the herpetology of Borneo. *Sarawak Mus. J.* 3(8): 15–34.

SMITH, M. A. 1926. Spolia Mentawia: Reptiles and amphibians. *Ann. & Mag. nat. Hist. Ser. 9* 18: 76–81.

SMITH, M. A. 1926. Monograph of the sea snakes (Hydrophiidae). British Museum of Natural History, London. xvii + (1) + 130 + (1) pp; 2 pl. Reprinted 1964, Wheldon and Wesley, Codicote.

SMITH, M. A. 1927. Contributions to the herpetology of the Indo-Australian region. *Proc. Zool. Soc., London* 1927: 199–225.

SMITH, M. A. 1930. The Reptilia and Amphibia of the Malay Peninsula from the Isthmus of Kra to Singapore including the adjacent islands. *Bull. Raffles Mus.* 3: (2) + xviii + 1–149.

SMITH, M. A. 1931. The herpetology of Mt. Kinabalu, North Borneo, 13,455 ft. *Bull. Raffles Mus.* 5: 3–32.

SMITH, M. A. 1931. Fauna of British India, including Ceylon and Burma. Vol. I. Loricata, Testudines. Taylor and Francis, London. xxviii + 185 pp + Pl.1–2.

SMITH, M. A. 1935. The sea snakes (Hydrophiidae). *DANA Rep.* (8): 1–6.

SMITH, M. A. 1935. Fauna of British India, including Ceylon and Burma. Vol. II. Sauria. Taylor and Francis, London. xiii + 440 pp + Pl. 1 + 2 maps.

SMITH, M. A. 1937. *Draco blanfordi* and its allies. *Bull. Raffles Mus.* 13: 75–79; Pl. VIII.

SMITH, M. A. 1937. A review of the genus *Lygosoma* (Scincidae: Reptilia) and its allies. *Rec. Indian Mus.* 39(3): 213–234.

SMITH, M. A. 1938. The nucho-dorsal glands of snakes. *Proc. Zool. Soc. London Ser. B* 107: 575–583.

SMITH, M. A. 1940. The herpetology of the Andaman and Nicobar Islands. *Proc. Linn. Soc.* 1940: 150–158.

SMITH, M. A. 1941–42. The biogeographic division of the Indo-Australian archipelago 3—The divisions as indicated by the Vertebrata. *Proc. Linn. Soc, London* 1941–42: 138–142.

SMITH, M. A. 1943. Fauna of British India, Ceylon and Burma, including the whole of the Indo-Chinese region. Vol. III. Serpentes. Taylor and Francis, London. xii + 583 pp.

SMITH, M. A. & J. B. PROCTER. 1921. On a collection of reptiles and batrachians from the island of Ceram, Indo-Australian Archipelago. *Ann. & Mag. nat. Hist. Ser. 9* 7: 352–355.

SMITH, T. 1988. *Gonyosoma oxycephala* the red-tailed racer. Asiatic colubrids, Part 4. *The Herpetile* 13(4): 154–170.

SOCHUREK, E. & F. J. OBST. 1986. *Elaphe porphyracea*, Rote Bambusnatter. *Elaphe* 2(4): 80.

SODEN, S. & M. WHYTE. 1996. Breeding the Asian yellow-spotted climbing toad. *Reptiles* 4(4): 68–73.

SOMANTRI, A. & (I.) S. SUWELO. 1986. Sea turtles in Indonesia: economic prospects and conservation. *Mar. Turtle Newsl.* (38): 3.

SONDAAR, P. Y. 1981. The *Geochelone* faunas of the Indonesian archipelago and their paleogeographical and biostratigraphical significance. *Mod. Quart. Res. Southeast Asia* 6: 111–120.

SONINI, C. S. & P. A. LATREILLE. 1801. Histoire naturelle des reptiles, avec figures dessinées d'après nature. Deterville, Paris. 335 pp.

SONNEMANN-REBENTISCH, J. H. A. B. 1959. Reptiliën van Borneo. *Natuur. Tijd. Ned-Indië* 16(2): 37–38.

SONNEMANN-REBENTISCH, J. H. A. B. 1959. Reptiliën van Borneo. *Natuur. Tijd. Ned-Indië* 16(2): 188.

SONNEMANN-REBENTISCH, J. H. A. B. & P. BLEEKER. 1959. Verslag van reptiliën en visschen van Westelijk Borneo. *Natuur. Tijd. Ned-Indië* 16(2): 435–441.

SPRACKLAND, R. G. 1970. Further notes on *Lanthonotus. Sarawak Mus. J.* 18: 412–413.

SPRACKLAND, R. G. 1972. A summary of observations of the earless monitor, *Lanthonotus borneensis. Sarawak Mus. J.* 20: 323–327.

SPRACKLAND, R. G. 1976. Notes on Dumeril's monitor lizard, *Varanus dumerili* (Schlegel). *Sarawak Mus. J.* 24: 287–291.

SPRACKLAND, R. G. 1993. The taxonomic status of the monitor lizard, *Varanus dumerilii heteropholis* Boulenger, 1892 (Reptilia: Varanidae). *Sarawak Mus. J. n.s.* 44(65): 113–121.

STEINDACHNER, F. 1867. Zoologischer Theil, Band 1. Reptilien. *In:*

Reise der österreichischen Fregatte Novara um die Erde in der Jahren 1857, 1858, 1859 unter den Befehlen des Commodore B. von Wüllerstorf-Urbair. Kaiserlich-Königlischen Hof-und Staatsdruckerei, Wien. 98 pp + Pl. I–III.

STEJNEGER, L. H. 1902. *Ptychozoon kuhli* a new name for *P. homalocephalum. Proc. Biol. Soc. Washington* 15: 37.

STEJNEGER, L. H. 1922. List of snakes collected in Bulungan, northeast Borneo by Cark Lumholtz, 1914. *Nyt Magazin for Naturvidensk* 60(2): 77–84.

STEJNEGER, L. H. 1933. Crocodilian nomenclature. *Copeia* 1933(3): 117–120.

STEJNEGER, L. H. 1933. The ophidian generic names *Ahaetulla* and *Dendrophis. Copeia* 1933(4): 199–203.

STEJNEGER, L. H. 1936. Types of the amphibian and reptilian genera proposed by Laurenti in 1768. *Copeia* 1936(3): 133–141.

STEJNEGER, L. H. 1940. Status of the name *Rana grunniens* Daudin. *Copeia* 1940(3): 149–151.

STIMSON, A. F. 1969. Liste der rezenten Amphibien und Reptilien. Boidae (Boinae + Bolyeriinae + Loxoceminae + Pythoninae). Das Tierreich 89. Walter de Gruyter & Co. I–XI + 1–49 pp.

STIMSON, A. F., J. ROBB & G. UNDERWOOD. 1977. *Leptotyphlops* and *Ramphotyphlops* Fitzinger, 1843 (Reptilia, Serpentes): proposed conservation under the plenary powers. *Bull. Zool. Nom.* 33(3/4): 204–207.

STOLICZKA, F. 1870. Observations on some Indian and Malayan Amphibia and Reptilia. *Proc. Asiatic Soc. Bengal* 1870(4): 103–109

STOLICZKA, F. 1870. Observations on some Indian and Malayan Amphibia and Reptilia. *J. Asiatic Soc. Bengal* 39(2): 134–157 + captions, Pl. IX.

STOLICZKA, F. 1870. Observations on some Indian and Malayan Amphibia and Reptilia. *J. Asiatic Soc. Bengal* 39(3): 159–228, Pl. X–XII.

STOLICZKA, F. 1870. Observations on some Indian and Malayan Amphibia and Reptilia. *Ann & Mag. nat. Hist. ser.* 46(3): 105–109.

STOLICZKA, F. 1872. Notes on various new or little known Indian lizards. *J. Asiatic Soc. Bengal* 41(1): 86–116; Pl. II–V.

STOLICZKA, F. 1873. Notes on some Andamanese and Nicobarese reptiles. *Proc. Asiatic Soc. Bengal* (5): 118.

STOLICZKA, F. 1873. Note on some Andamanese and Nicobarese reptiles with descriptions of three species of lizards. *J. Asiatic Soc. Bengal* 42(3): 162–169.

STUEBING, R. B. 1983. Sarawak's killer crocodiles. *Malay. nat. J.* 37(1): 20–30.

STUEBING, R. B. 1984. Creeping frogs of Sabah. *Sabah Soc. J.* 7(4): 299–301.

STUEBING, R. B. 1985. Batang Lupar crocodiles: Happy bachelors, lovesick behemoths, or nasty brutes? *Malayan Natural.* 39(1/2): 43–46.

STUEBING, R. B. 1986. The Kinabalu tree-frogs. *Sabah Soc. J.* 8(2): 210–212.

STUEBING, R. B. 1988. Island romance: the biology of the yellow-lipped sea krait. *Malayan Natural.* 41: 9–11.

STUEBING, R. B. 1989. Notes on the riparian frogs of the Tabin Wildlife Reserve, Sabah. *Sabah Mus. Monogr.* 3: 100–103.

STUEBING, R. B. 1991. A checklist of the snakes of Borneo. *Raffles Bull. Zool.* 39: 323–362.

STUEBING, R. B. 1994. A checklist of the snakes of Borneo: addenda and corrigenda. *Raffles Bull. Zool.* 42(4): 931–936

STUEBING, R. B. 1994. A new species of *Cylindrophis* (Serpentes: Cylindrophiidae) from Sarawak, western Borneo. *Raffles Bull. Zool.* 42(4): 967–973.

STUEBING, R. B. 1996. The herpetofauna of Lanjak-Entimau Wildlife Sanctuary. Seminar on the Development and Management of the Lanjak-Entimau Biodiversity Conservation Area. Forest Department Sarawak/International Tropical Timber Organization, Kuching. 16 pg.

STUEBING, R. B. & R. GOH. 1993. A new record of Leonard's pipe snake, *Anomochilus leonardi* Smith (Serpentes: Uropeltidae: Cylindrophiinae) from Sabah, northwestern Borneo. *Raffles Bull. Zool.* 41(2): 311–314.

STEUBING, R. B., G. ISMAIL & H. C. LING. 1992. Distribution and abundance of the Indo-Pacific crocodile (*Crocodylus porosus* Schneider) in the Klias River, Sabah. *In*: Forest biology and conservation in Borneo. pp: 492–493. G. Ismail, M. Mohamed & S. Omar (Eds.). Centre for Borneo Studies Publication No. 2. Yayasan Sabah, Kota Kinabalu.

STUEBING, R. B., E. LADING & F. S. P. LIEW. 1990. Snake Island of Pulau Tiga Park. Sabah Parks Trustees, Kota Kinabalu. (4) + 29 pp.

STUEBING, R. & H. K. VORIS. 1990. Relative abundance of marine snakes on the west coast of Sabah, Malaysia. *J. Herpetol.* 24(2): 201–202.

STULL, O. G. 1935. A check list of the family Boidae. *Proc. Boston Soc. nat. Hist.* 40(8): 387–408.

SUAREZ, A. & C. H. STARBIRD. 1995. A traditional leatherback fishery in Maluku. *Mar. Turtle Newsl.* (68): 15–18.

SUAREZ, A. & C. H. STARBIRD. 1996. Subsistence hunting of leatherback turtles, *Dermochelys coriacea*, in the Kai Islands, Indonesia. *Chelonian Conserv. & Biol.* 2(2): 190–195

SUDHARMA, D. 1976. [Some aspects of the life history of the bowspirit crocodile (*Tomostoma schlegelii*) and the possible building up of population.] Department Biologi Perairan, Fakultas Perikanan, Institut Pertanian, Bogor. [In Bahasa Indonesia.]

SUMERTHA, I. & K. SUKARNA. 1977. Mungkinkah terdapt penyu laut lebih dari dua jenis di perairan Sumatra Barat. Institute Pertanian, Fakultas Perikanan, Bogor. 19 pg.

SURAHYA, S. [An anatomical study of the Komodo dragon and its position in animal systematics.] Gadjah Madah University Press, Yogyakarta. Vol. 1, xix + 324 pp; Vol. 2, 382 colour photos. [In Bahasa Indonesia.]

SUWELO, I. S. 1971. Sea turtles in Indonesia. *IUCN Publ. (n.s.) Suppl. Pap.* (31): 85–89.

SUWELO, I. S. 1990. Indonesian hawksbill turtle ranching: a pilot project. *Mar. Turtle Newsl.* (49): 16.

SUWELO, I. S. 1992. Hawksbill turtles in Seribu Islands, Jakarta Bay, Indonesia. *Mar. Turtle Newsl.* (59): 4.

SUWELO, I. S., N. S. SUITJA & I. SOETRISNO. 1982. Marine turtles in Indonesia. *In*: Biology and conservation of sea turtles. pp: 349–351. K. A. Bjorndal (Ed.). Smithsonian Institution Press, Washington, D.C.

SUZUKI, A. 1996. *Megophrys nasuta* Schlegel 1837. *Herpetomania* 5(6): 11–15.

TAN, C. K. & P. GOPALAKRISHNAKONE. 1987. Some observations on a unique longitudinal muscle band in the stomach of the python (*Python reticulatus*). *The Snake* 19(1): 41–46.

TAN, F. L. 1988. Survey report on frogs of Kinabalu Park. Report to Sabak Parks, Kota Kinabalu. 27 pg.

TAN, F. L. 1992. Ecological distribution of amphibians and reptiles in lowlands of Crocker Range National Park, Sabah. *In*: Forest biology and conservation in Borneo. pp: 496–497. G. Ismail, M. Mohamed & S. Omar (Eds.). Centre for Borneo Studies Publication No. 2. Yayasan Sabah, Kota Kinabalu.

TANZER, S. L. & J. H. R. W. C. VAN HEURN. 1938. Observations made by S. L. Tanzer and J. H. R. W. C. van Heurn with reference to the propagation of the *Varanus komodoensis* Ouw. *Treubia* 16(3): 365–368.

TAYLOR, E. H. 1920a. Philippine turtles. *Philippines J. Sci.* 16: 111–144.

TAYLOR, E. H. 1920b. Philippine Amphibia. *Philippines J. Sci.* 16: 213–359.

TAYLOR, E. H. 1921. Amphibians and turtles of the Philippines. *Philippines J. Sci.* 15: 1–193.

TAYLOR, E. H. 1922. The snakes of the Philippine Islands. Philippine Bureau of Science, Manila. 312 pp; 37 pl.

TAYLOR, E. H. 1922. The lizards of the Philippine Islands. Philippine Bureau of Science, Manila. 269 pp + 23 pl.

TAYLOR, E. H. 1925. Additions to the herpetological fauna of the Philippines. IV. *Philippine J. Sci. D* 26: 97–111.

TAYLOR, E. H. 1928. Amphibians, lizards, and snakes of the Philippines. *In*: Distribution of life in the Philippines. R. Dickerson (Ed.). *Bur. Sci. Monogr.* (21): 214–241; Pls. 27–32.

TAYLOR, E. H. 1965. New Asiatic and African caecilians with redescriptions of certain other species. *Univ. Kansas Sci. Bull.* 46(6): 253–302.

TAYLOR, E. H. 1968. The caecilians of the world: A taxonomic review. University of Kansas Press, Lawrence. XIV + 848 pp.

TIEDEMANN, F., M. HÄUPL & H. GRILLITSCH. 1994. Kataloge der wissenschaftlichen Sammlungen des Naturhistorischen Museums in Wien. Vertebrata Heft 4. Katalog der Typen der Herpetologischen Sammlung nach dem Stand vom 1. Jänner 1994. Band 10. Teil II: Reptilia. Naturhistorisches Museum Wien, Vienna. 110 pp.

TIKADER, B. K., A. DANIEL & N. V. SUBBA RAO. 1986. Sea shore animals of Andaman & Nicobar Islands. Zoological Survey of India, Calcutta. xii + 188 pp.

TIKADER, B. K. & A. K. DAS. 1985. Glimpses of animal life of Andaman & Nicobar Islands. Zoological Survey of India, Calcutta. xi + 170 pp.

TIMMERMAN, W. W. & D. L. AUTH. 1988. *Heosemys leytensis* (Leyte pond turtle): Philippines: Palawan Province. *Herpetol. Rev.* 19(1): 21.

TIWARI, K. K. 1961. The eggs and flight of gecko *Ptychozoon kuhli* Stejneger from Car Nicobar. *J. Bombay nat. Hist. Soc.* 58: 523–526.

TIWARI, K. K. 1973. Two new reptiles from the Great Nicobar Island. *J. Zool. Soc. India* 25: 57–63.

TIWARI, M. 1991. A follow-up sea turtle survey in the southern Nicobars. Report to the Andaman and Nicobar Islands Forest Department and Madras Crocodile Bank Trust. 20 pg.

TIWARI, M. 1992. First record of the sunbeam snake *Xenopeltis unicolor* Reinwardt, 1827 (Serpentes: Xenopeltidae) from Great Nicobar Island. *J. Bombay nat. Hist. Soc.* 89: 383.

TIWARI, M. 1996. Les tortues de la mer des iles d'Andamans et Nicobars. *In*: International Congress of Chelonian Conservation. Proceedings. pp: 50–51. SOPTOM (Ed.). Editions Soptom, Gonfaron.

TORIBA, M. 1989. Karyotypes of two species of elapid snakes. *The Snake* 21(2): 87–89.

TORIBA, M. 1992. A comment on the status of the Thai palm viper, *Trimeresurus wiroti*. *The Snake* 24(1): 71–73.

TORIBA, M. 1992. Karyotype of red-tailed pipe snake, *Cylindrophis rufus* (Uropeltidae: Cylindrophiinae). *The Snake* 24(2): 143–146.

TORIBA, M. 1994. Gender of the genus *Amphiesma* Dumeril, Bibron and Dumeril. *The Snake* 26(2): 145. [In Japanese, with English abstract.]

TU, A. T. & S. GANTHAVORN. 1978. Comparison of snake venoms (Reptilia: Serpentes) from Java, Indonesia and Thailand and its significance and zoogeography. *J. Herpetol.* 12: 105–107.

TUBB, J. A. 1949. A "flying" snake. *Sarawak Mus. J.* 5: 153.

TWEEDIE, M. W. F. 1949. The flying gecko, *Ptychozoon kuhli* Stejn. *Proc. Zool. Soc. London* 120(1): 13.

TWEEDIE, M. W. F. 1949. Reptiles of the Kelabit Plateau. *Sarawak Mus. J.* 5: 154–155.

TWEEDIE, M. W. 1952. The paradise tree snake. *Malay. nat. J.* 7(2): 67.

TWEEDIE, M. W. F. 1953. The breeding of the leathery turtle. *Proc. Zool. Soc. London* 123(2): 273–275.

TWEEDIE, M. W. F. 1983. The snakes of Malaya. Third edition. Singapore National Printers (Pte) Ltd., Singapore. 167 pp.

TWEEDIE, M. W. F. 1990. Poisonous animals of Malaysia. Graham Brash, Singapore. (4) + 78 pp.

TWEEDIE, M. W. F. & H. A. REID. 1956. Poisonous snakes in Malaya. Malayan Museum Popular Pamphlet No. 11. 16 pp.

TYTLER, L. C. R. C. 1864. Observations on a few species of geckos alive in the possession of the author. *J. Asiatic. Soc. Bengal* 33: 535–548.

UCHIDA, I. 1979. [Brief report on the hawksbill turtle (*Eretmochelys imbricata*) in the waters adjacent to Indonesia and Malaysia.] Japanese Tortoise Shell Society, Nagasaki. [In Japanese.]

UCHIDA, I. 1980. [The report of a feasibility research on artificial hatchery and cultivation of hawksbill turtle *Eretmochelys imbricata* in waters adjacent to Malaysia, Singapore and Indonesia.] Japanese Tortoise Shell Society, Nagasaki. [In Japanese.]

ULBER, T. & E. ULBER. 1991. *Cosymbotus platyurus* (Schneider). *Sauria, Berlin* Suppl. 13(1–4): 201–204.

UNDERWOOD, G. 1957. *Lanthonotus* and the anguinomorphan lizards: A critical review. *Copeia* 1957(1): 20–30.

UNDERWOOD, G. 1967. A contribution to the classification of snakes. British Musuem (Natural History), London. x + 179 pp.

UNDERWOOD, G. "1997" 1996. An overview of venomous snake evolution. *In*: Venomous snakes: ecology, evolution and snakebite. pp: 1–13. R. S. Thorpe, W. Wüster & A. Malhotra (Eds.). Symp. Zool. Soc. London (70). The Zoological Society of London/Clarendon Press, Oxford.

UNDERWOOD, G. & A. F. STIMSON. 1990. A classification of pythons (Serpentes, Pythoninae). *J. Zool. London* 221: 565–603.

USSHER, C. A. 1979. Brunei's largest snake. *Brunei Mus. J.* 4(3): 180–181.

"V." 1924. Iets over de oelar Bakas (*Python curtus*). *Trop. Natuur* 13: 109–110.

VALLE, G. H. 1994. Quintessential Asian geckos. *Reptile & Amphibian Mag.* Nov./Dec.: 33–41.

VAN BLOEMEN, J. T. 1859. Reptiliën en visschen van Bali. *Natuur. Tijd. Ned-Indië* 16(2): 267.

VAN BLOEMEN, J. T. 1860. Reptiliën van Boni. *Natuur. Tijd. Ned-Indië* 22(2): 81–83.

VAN BLOEMEN, J. T., F. W. SYTHOFF & P. BLEEKER. 1859. Verslang omtrent vischsoorten en reptiliën van Ngawi. *Natuur. Tijd. Ned-Indië* 16(2): 358–359.

VAN DE BUNT, P. 1990. A note on the exploitation of sea turtles in Sumatra. *Mar. Turtle Newsl.* (51): 19–20.

VAN DE BUNT, P. 1990. Tortoise exploitation in Sumatra. *Turtles & Tortoises, IUCN Tortoise & Freshwater Turtle Sp. Gr.* 5: 14–16.

VANDELLI, A. 1833. *In*: Reptiles et Poissons (of the Moreau). G. Bibron & J. B. G. M. Bory de Saint-Vincent (authors). Fr. Expéd. Sci. Moree 3(1): 1–209.

VAN DER HASSELT, F. J. F. 1922. Over het eten van schildpadvleisch. *Trop. Natuur* 11: 157–159.

VAN DER MEER MOHR, J. C. 1926. Aanteekeningen over *Python reticulatus. Trop. Natuur* 15: 9–14.

VAN DER MEER MOHR, J. C. 1927. Aantekeningen betreffende de biologie van *Chelonia mydas. Trop. Natuur* 16: 44–53.

VAN DER MEER MOHR, J. C. 1927. Bestaan er vliegende slangen. *Trop. Natuur* 16: 106–107.

VAN DER MEER MOHR, J. C. 1927. Het slangennest in een bloempot (*Macropisthodon flaviceps*). *Trop. Natuur* 16: 195.

VAN DER MEER MOHR, J. C. 1927. Notiz über seeschlangen. *Misc. Zool. Sumatrana* (23): 1–2.

VAN DER MEER MOHR, J. C. 1930. Over eieren von *Varanus salvator* en *Python curtus. Trop. Natuur* 19: 156–157.

VAN DER MEER MOHR, J. C. 1932. Ueber Anomalien in der Plattenzahl des Ruckenschildes bei *Callagur borneensis. Misc. Zool. Sumatrana* (6): 2.

VAN DER MEER MOHR, J. C. 1933. Enkele gegevens over *Callagur borneensis* (Schleg. et Müll.). *Trop. Natuur* 22: 29–31.

VAN DER MEER MOHR, J. C. 1934. Korte zoologische aanteekeningen II. *Callagur borneensis* (Schleg. & Müll.). *Trop. Natuur* 23: 193.

VAN DIJK, A. 1935. Ausfuhrgunge der Leben und das Pankreas bei *Varanus komodoensis* und einigen anderen Reptilien. *Anat. Anz.* 6(17/20): 347–355.

VAN EIJSDEN, E. H. T. 1976. Vroedmeester bij *Cosymbotus platyurus* (Schneider). *Lacerta* 34(4): 35–36.

VAN EIJSDEN, E. H. T. 1982. *Hemidactylus frenatus*, de tjitjak. *Lacerta* 40(10/11): 221–222.

VAN HASSELT, A. W. M. 1858. Aanteekening over en nadere beschrijving van een individu der grootste tot nu bekendre gift-slangen uit het geslacht der *Naja's. Versl. Med. Koninkl. Akad. Wetensch. (Natuurk.) ser. 1* 7: 200–208.

VAN HASSELT, A. W. M. 1882. Eene monster-*Naja. Versl. Med. Koninkl. Akad. Wetensch. (Natuurk.) ser. 2* 17: 140–143.

VAN HEURN, W. C. 1929. *Coluber radiatus*, de "Tjaoe asak". *Trop. Natuur* 18: 128–130.

VAN HEURN, W. C. 1929. *Lachesis puniceus* Boie. *Trop. Natuur* 18: 171–175.

VAN HEURN, W. C. 1930. Herpetologische aanteekeningen, I. *Bungarus fasciatus, Doliophis bivirgatus. Trop. Natuur* 19: 123–125.

VAN HEURN, W. C. 1930. Herpetologische aanteekeningen, II. *Trop. Natuur* 19: 174–176.

VAN HEURN, W. C. 1930. Herpetologische aanteekeningen, III. *Doliophis intestinalis. Trop. Natuur* 20: 61–63.

VAN HEURN, W. C. 1931. *Bufo asper. Trop. Natuur* 20: 73.

VAN HEURN, W. C. 1931. *Coluber* (= *Elaphe*) *porphyracea* Boulenger. *Trop. Natuur* 20: 201.

VAN HEURN, W. C. 1930. Herpetologische aanteekeningen, IV. *Naja tripudians sputatrix. Trop. Natuur* 20: 205–209.

VAN HOESEL, J. K. P. 1948. Snake hunting. *Treubia* 19(3): 525–538.

VAN HOESEL, J. K. P. 1954. *Vipera russelli*—its zoogeographic range and local distribution in Indonesia. *Trop. Natuur* 33: 133–139.

VAN HOESEL, J. K. P. 1958. Notities over *Vipera russellii* en enkele andere slangen van Flores. *Lacerta* 16(1): 32–36.

VAN HOESEL, J. K. P. 1959. Ophidia Javanica. Museum Zoologicum Bogoriense, Bogor. 188 pp.

VAN KAMPEN, P. N. 1905. Amphibien von Palembang (Sumatra). *Zool. Jahrb., (Syst)* 22: 701–716; Pl. 26.

VAN KAMPEN, P. N. 1907. Amphibien des indischen Archipels. *In*: Zoologische ergebnisse einer reise in Niederlandisch Ost-Indien, herausgegeben von Dr. Max Weber 4(2): 383–417; Pl. 16. E. J. Brill, Leiden.

VAN KAMPEN, P. N. 1909. Liste der Amphibiens des Indischen Archipels im Museum zu Buitenzorg. *Bull. Dept. Agric.* 25: 3–9.

VAN KAMPEN, P. N. 1910. Eine neue *Nectophryne*-Art und andere Amphibien von Deli (Sumatra). *Nat. Tijd. Ned.-Indië* 69: 18–24; 1 pl.

VAN KAMPEN, P. N. 1910. Beitrag zur kenntnis der Amphibienlarven des indischen Archipels. *Nat. Tijd. Ned.-Indië* 69: 25–48; 1 pl.

VAN KAMPEN, P. N. 1913. Amphibien von Waigen un den Molukken. *Bijdragen tot de Dierkunde* 19: 89–92.

VAN KAMPEN, P. N. 1923. The Amphibia of the Indo-Australian Archipelago. E. J. Brill, Leiden. xii + 208 pp.

VAN KAMPEN, P. N. & L. D. BRONGERSMA. 1931. Notes on a small collection of Amphibia from Sumba. *Treubia* 13(1): 15–18.

VAN LIDTH DE JEUDE, T. W. 1886. Note X. On *Cophias wagleri*, Boie and *Coluber sumatranus*, Raffles. *Notes Leyden Mus.* 8: 43–54; Pl. 2

VAN LIDTH DE JEUDE, T. W. 1890. Reptilia from the Malay Archipelago. II. Ophidia. *In*: Zoologische Ergebnisse einer Reise in Niederländisch ost-Indien 1: 178–192; Pl. XV–XVI. M. Weber (Ed.). E. J. Brill, Leiden.

VAN LIDTH DE JEUDE, T. W. 1890. On a collection of reptiles from Deli. *Notes Leyden Mus.* 12(1): 17–27.

VAN LIDTH DE JEUDE, T. W. 1890. On a collection of reptiles from Nias, and on *Calamaria virgulata*, Boie. *Notes Leyden Mus.* 12(4): 253–256.

VAN LIDTH DE JEUDE, T. W. 1893. On reptiles from North Borneo. *Notes Leyden Mus.* 15: 250–257.

VAN LIDTH DE JEUDE, T. W. 1895. Reptiles from Timor and the neighbouring islands. *Notes Leyden Mus.* 16: 119–127.

VAN LIDTH DE JEUDE, T. W. 1896. On *Testudo emys* Schleg. & Müll. and its affinities. *Notes Leyden Mus.* 17: 197–204; Pl. 5–6.

VAN LIDTH DE JEUDE, T. W. 1904. Reptilia from the Malay Archipelago. *In*: Zoologische Ergebnisse einer Reise in Niederländisch ost-Indien 2: 178–192. M. Weber (Ed.). E. J. Brill, Leiden.

VAN LIDTH DE JEUDE, T. W. 1905. Zoological results of the Dutch Scientific Expedition to Central Borneo. The reptiles. Part 1. Lizards. *Notes Leyden Mus.* 25: 187–202.

VAN LIDTH DE JEUDE, T. W. 1922. Snakes from Sumatra. *Zool. Med.* 6: 239–253.

VAN LIDTH DE JEUDE, T. W. 1930. A new *Amblycephalus* from Sumatra. *Zool. Med.* 7: 243.

VAN RIEL, C. A. P. 1977. Kweekresultaat met *Boiga dendrophila melanota*. *Lacerta* (12): 182–183.

VAN RIEL, C. A. P. 1978. Élevage et reproduction en captivité de *Gonyosoma oxycephala* Boie. *Cobra* 4(18/19): 9–12.

VAN RIEL, C. A. P. 1978. Voortplanting in het terrarium van *Gonyosoma oxycephala* (Boie). *Lacerta* 37(2): 19–22.

VAN RIEL, C. A. P. 1991. Voortplanting in het terrarium van de Roodstaartslang (*Gonyosoma oxycephala*). *Lacerta* 50(1): 39–42.

VAN ROSENBERG, H. 1859. Reptilien van Amboina. *Natuur. Tijd. Ned-Indië* 16(2): 423.

VERHAART, P. 1976. Ervaringen met *Maticora intestinalis*. *Lacerta* 35(3): 31–35.

VERKERK, J. W. 1983. Opmerkelijk gedrag van de vliegende gekko, *Ptychozoon kuhli*. *Lacerta* 41(9): 176–177.

VETTER, R. S. & E. D. BRODIE. 1977. Background color selection and antipredator behavior of the flying gecko, *Ptychozoon kuhli*. *Herpetologica* 33(4): 464–467.

VINCIGUERRA, D. 1892. Rettili e batrici di Engano raccolti dal Dott. Elio Modigliani. *Ann. Mus. Genova Ser. 2* 12(32): 517–526.

VOGEL, G. 1995. *Dendrelaphis striatus* (Cohn), neu für die Fauna Borneos (Serpentes, Colubridae). *Mitt. Zool. Mus. Berlin* 71: 147–149.

VOGEL, G. & W. GROSSMANN. 1993. *Chrysopelea pelias* (Linnaeus). *Sauria, Berlin* 15(1): 2.

VOGEL, G. & P. HOFFMANN. 1997. Über die Lebenfarbung von *Bungarus flaviceps baluensis* Loveridge, 1938 (Serpentes: Elapidae). *Sauria, Berlin* 19(1): 13–16.

VOGEL, P. 1979. Zur Biologie des Bindenwarans (*Varanus salvator*) im westjavanischen naturschultzgebiet Ujung Kulon. Ph. D. Dissertation, University of Basel. 139 pg.

VOGEL, P. 1979. Innerartliche Anseinandersetzungen bei freilebenden Bindenwaranen (*Varanus salvator*). *Salamandra* 15(2): 65–83.

VOGT, T. 1925. Beitrag zur Kenntnis der Schlangengattung *Calamaria*. *Zool. Anz.* 62: 64–65.

VOLZ, W. 1903. Lacertilia von Palembang (Sumatra). *Zool. Jb. Abt. f. Syst.* 19: 421–430.

VOLZ, W. 1904. Schlangen von Palembang (Sumatra). *Zool. Jb. Abt. f. Syst.* 20: 491–508.

VON TSCHUDI, J. J. 1828. Classification der Batrachier, mit Berucksichtigang der fossilen Thiere dieser Abtheilung der Reptilien. Petitpierre, Neuchâtel. 99 + (2) pp + 6 pl.

VORIS, H. K. 1964. Notes on the sea snakes of Sabah. *Sabah Soc. J.* 2(3): 138–141.

VORIS, H. K. 1977. Comparison of herpetofaunal diversity in tree buttresses of evergreen tropical forests. *Herpetologica* 33(3): 375–380.

VORIS, H. K. & D. R. KARNS. 1996. Habitat utilization, movements, and activity patterns of *Enhydris plumbea* (Serpentes: Homalopsinae) in a rice paddy wetland in Borneo. *Herpetol. Nat. Hist.* 4(2): 111–126.

VORSTMAN, A. G. 1928. Varanen en pythons. *Trop. Natuur* 17: 30–31.

VORSTMAN, A. G. 1933. The septa in the ventricle of the heart of *Varanus komodoensis. Proc. Koninkl. Akad. Wet.* 36(10): 911–913.

WAGLER, J. 1830. Natürliches System der Amphibien, mit vorangehender Classification der Saügthiere und Vögel. Ein Beitrag zur vergleichenden Zoologie. J. G. Cotta'schen Buchhandlung, München (= Munich), Stuttgart and Tubingen. vi + 354 pp.

WAGLER, J. 1828. Descriptiones et icones amphibiorum. Fascicle I. J. G. Cotta, München (= Munich). Unpaginated text + 12 pl.

WALL, F. 1914. Occurrence of Cantor's water snake in the Andamans. *J. Bombay nat. Hist. Soc.* 23: 166.

WALL, F. 1921. Remarks on a specimen of *Calamaria javanica. Rec. Indian Mus.* 22: 729.

WALLACE, A. R. 1860. On the zoological geography of the Malay archipelago. *J. Proc. Linn. Soc.* 14: 173–184.

WALLACE, A. R. 1869. The Malay Archipelago: the land of the orangutan and the bird of paradise. Oxford University Press, Oxford. 638 pp.

WALLACH, V. 1988. Status and redescription of the genus *Pandangia* Werner, with comparative visceral data on *Collorhabdium* Smedley and other genera (Serpentes: Colubridae). *Amphibia-Reptilia* 9: 61–76.

WALLACH, V. & A. M. BAUER. 1996. On the identity and status of *Simotes semicinctus* Peters, 1862 (Serpentes: Colubridae). *Hamadryad* 21: 13–18.

WALLIN, L. 1985. A survey of Linneaus's material of *Chelone mydas, Caretta caretta* and *Eretmochelys imbricata* (Reptilia, Cheloniidae). *Zool. J. Linn. Soc.* 85: 121–130.

WALSH, T. & R. ROSSCOE. 1994. The history, husbandry, and breeding of Komodo monitors at the National Zoological Park. pp: 276. *In*: Abstracts, 2nd World Congress of Herpetology. Adelaide, Australia

WANDOLLECK, B. 1900. Zur Kenntnis der Gattung *Draco. Abh. Ber. Mus. Tierk. Völkerk. Dresden* 9(3): 1–16; 1 pl.

WASSINK, G. 1860. Reptiliën van het eiland Timor. *Natuur. Tijd. Ned-Indië* 22(2): 86–88.

WEBB, R. G. 1995. Redescription and neotype designation of *Pelochelys bibroni* from southern New Guinea (Testudines: Trionychidae). *Chelonian Conserv. & Biol.* 1: 301–310.

WEBB, R. G. 1997. Geographic variation in the giant softshell turtle, *Pelochelys bibroni.* Linneaus Fund Research Report. *Chelonian Conserv. & Biol.* 2(3): 450.

WEBER, M. 1890. Reptilia from the Malay Archipelago. 1. Sauria, Crocodylidae, Chelonia. *In*: Zoologische Ergebnisse einer Reise in Niederländisch ost-Indien 1: 158–177. M. Weber (Ed.). E. J. Brill, Leiden.

WEGNER, A. M. R. 1954. The snakes of the genus *Maticora* Gray with special reference to *Maticora bivirgata* (Boie) and *M. intestinalis intestinalis* (Laurenti). *Peng. Alam* 34(1–2): 55–58.

WEIGMANN, A. F. A. 1834. Beiträg zur Zoologie, gesammelt auf einer Reise um die Erde, von Dr. F. J. F. Meyer, M. d. A. d. N. Siebente Abhandlung. Amphibien. *Nova Acta Acad. Caes. Leopold Carol., Halle* 17(1): 183–268; 268a–d; 10 pl.

WELCH, K. R. G. 1988. Snakes of the Orient: A checklist. Robert E. Krieger Publishing Company, Malabar, Florida. viii + 183 pp.

WELCH, K. R. G., P. S. COOKE & A. S. WRIGHT. 1990. Lizards of the Orient: A checklist. Robert E. Krieger Publishing Company, Malabar, Florida. v + 162 pp.

WERMUTH, H. 1954. Zur Nomenklatur und Typologie des Leistenkrokodils, *Crocodilus porosus* Schneider 1801. *Mitt. Zool. Mus. Berlin* 30: 483–486; Pl. 63.

WERMUTH, H. 1966. Liste der rezenten Amphibien und Reptilien. Gekkonidae, Pygopodidae, Xanthusiidae. Das Tierreich 80. Walter de Gruyter & Co. I–XXII + 1–246 pp.

WERMUTH, H. 1967. Liste der rezenten Amphibien und Reptilien. Agamidae. Das Tierreich 86. Walter de Gruyter & Co. I–XIV + 1–127 pp.

WERMUTH, H. & R. MERTENS. 1977. Liste der rezenten Amphibien und Reptilien. Testudines, Crocodylia, Rhynchocephalia. Das Tierreich 100. Walter de Gruyter & Co. 1–174 pp.

WERNER, F. 1892. Ueber eine kleine Kollection von Reptilien und Batrachiern von Nias. *Jb. Nat. Ver. Magdeburg* 1892: 248–254.

WERNER, F. 1893. Bemerkungen über Reptilien und Batrachier aus dem tropischen Asien und von der Sinai-Halbinsel. *Verh. Zool. Bot. Ges. Wien* 43: 349–359.

WERNER, F. 1896. Zweiter Beitrag zur Herpetologie der indo-orientalischen Region. *Verh. Zool. Bot. Ges. Wien* 46: 6–24; 1 pl.

WERNER, F. 1900. Reptilien und Batrachier aus Sumatra, gesammelt von Herrn Gustav Schneider jr. im Jahre 1897–98. *Zool. Jahrb. Syst.* 13: 479–508; Pl. 31–35.

WERNER, F. 1909. Über neue oder seltene Reptilien des Naturhistorischen Museums in Hamburg. I. Schlangen. *Mitt. Naturhist. Mus. Hamburg* 26: 205–247.

WERNER, F. 1910. Neue oder seltnere Reptilien des Musée royal d'Histoire naturelle de Belgique in Brussel. *Zool. Jb. (Abt. Syst.)* 28: 263–288.

WERNER, F. 1913. Neue oder seltene Reptilien und Frösche des Naturhistorischen Museums Hamburg. *Mitt. Naturhist. Mus. Hamburg* 30: 1–51.

WERNER, F. 1917. Über einige neue Reptilien und einem neuen Frösch des Zoologischen Museums in Hamburg. *Jb. Hamburg Wiss. Anst.* 34: 31–36.

WERNER, F. 1921. Synopsis der Schlangenfamilie der Boiden auf Grundlage des Boulenger'schen Schlangenkatalogs (1893–1896). *Arch. Naturgesch. Berlin A* 87(7): 230–265.

WERNER, F. 1921. Synopsis der Schlangenfamilie der Typhlopiden auf

Grund des Boulenger'schen Schlangenkatalogs (1893–1896). *Arch. Naturgesch. Berlin A* 87(7): 266–330.

WERNER, F. 1922. Synopsis der Schlangenfamilien der Amblycephaliden und Viperiden nebst Übersicht über die kleineren Familien und die Colubriden der Acrochordinengruppe. Auf Grund des Boulenger'schen Schlangenkatalogs (1893–1896). *Arch. Naturgesch. Berlin A* 88(8): 185–244.

WERNER, F. 1923. Übersicht der Gattungen und Arten der Schlangen der Familie Colubridae. I. Teil. Mitt. einem Nachtrag zu den übrigen Familien. *Arch. Naturgesch. Berlin A* 89(8): 138–199.

WERNER, F. 1924. Neue oder wenig bekannte Schlangen aus dem Natur-historichen Staatmuseum in Wien. *Sitz. Akad. Wiss. Math.-Naturw. Klasse, Wien 1* 133: 29–56.

WERNER, F. 1925. Neue oder wenig bekannte Schlangen aus dem Natur-historichen Staatmuseum in Wien. *Sitz. Akad. Wiss. Math.-Naturw. Klasse, Wien 1* 134: 46–66.

WERNER, F. 1925. Übersicht der Gattungen und Arten der Schlangen der Famillie Colubridae. II Tiel. *Arch. Naturg. A* 90: 108–166.

WERNER, F. 1925. Neue oder wenig bekannte Schlangen auf dem Wiener naturhistorische Staatmuseum. 2. Tiel. *Sitzb. Akad. Wiss. Wien 1* 134: 45–66.

WERNER, F. 1927. Über Schlangen von Medan, Sumatra's Ostküste, (gesammel von Reg. Rat Dr. Leopold Fulmek). *Misc. Zool. Sumatrana* (19): 1–3.

WERNER, F. 1929. Übersicht der Gattungen und Arten der Schlangen aus der Familie Colubridae. III. Teil (Colubrinae). Mit einem Nachtrag zu den übrigen Familien. *Zool. Jahrb. Abt. Syst.* 57: 1–196.

WESTERMANN, J. H. 1942. Snakes from Bangka and Billiton. *Treubia* 18(3): 611–619.

WHITAKER, R. 1978. Herpetological survey in the Andamans. *Hamadryad* 3: 9–16.

WHITAKER, R. 1978. Birth record of the Andaman pit viper, *Trimeresurus purpureomaculatus*. *J. Bombay nat. Hist. Soc.* 75(1): 233.

WHITAKER, R. 1982. An Andaman experience. *Sanctuary Asia* 2(4): 316–325.

WHITAKER, R. 1983. Crocodile resources in the Andaman and Nicobars. Marine potential. *Bull. Cent. Mar. Fish. Res. Inst.* 34: 100–101.

WHITAKER, R. 1983. Preliminary survey of crocodiles in Sabah, East Malaysia. Report to the World Wildlife Fund-Malaysia and Game Branch, Sabah Forest Department. 12 pg. [Mimeo].

WHITAKER, R. 1984. Borneo crocodile survey. Part II. *Hamadryad* 9(1): 5–9.

WHITAKER, R. 1984. A preliminary survey of the crocodiles in Sabah, East Malaysia. Report to WWF-Malaysia, Kota Kinabalu. 35 pg.

WHITAKER, R. 1985. Rational use of estuarine and marine reptiles. *In*: Proc. Symp. Endangered Mar. Animals & Mar. Parks. 1: 298–303. Cent. Mar. Fish. Res. Inst. Spec. Publ. 18. CMFRI, Cochin.

WHITAKER, R. 1990. Why did they call it Snake Island? *Daily Telegrams, Port Blair* 16 May, 1990.

WHITAKER, R. 1991. Suspected case of death by pit viper bite. *Hamadryad* 16: 37.

WHITAKER, R. 1993. Eco-development and wildlife management in the Andaman and Nicobar Archipelago. *In*: Spirit of Enterprise: The 1993 Rolex Awards. pp: 332–334. D. W. Reed (Ed). Buri International, Bern.

WHITAKER, R. 1994. Some environmental issues facing the Andaman and Nicobar Islands. *In*: Andaman and Nicobar Islands—challenges for development. pp: 97–106. V. Suryanarayan & V. Sunderesan (Eds). Konark Pub., New Delhi.

WHITAKER, R. & Z. WHITAKER. 1978. A preliminary survey of the saltwater crocodile (*Crocodylus porosus*) in the Andaman Islands. *J. Bombay nat. Hist. Soc.* 76: 311–325.

WHITAKER, R. & Z. WHITAKER. 1978. Notes on *Phelsuma andamanense*, the Andaman day gecko or green gecko. *J. Bombay nat. Hist. Soc.* 75(2): 497–499.

WHITAKER, R. & Z. WHITAKER. 1989. The status and conservation of the Asian crocodilians. *In*: Crocodiles: their ecology, management and conservation. pp: 297–308. IUCN Special Publication, IUCN, Gland.

WHITTEN, A. J., S. J. DAMANIK, J. ANWAR & N. HISYAM. 1987. The ecology of Sumatra. Gadja Madah University Press, Yogyakarta. xxii + 583 pp.

WHITTEN. A. J. & C. McCARTHY. 1993. List of the amphibians and reptiles of Jawa and Bali. *Trop. Biodiv.* 1(3): 169–177.

WHITTEN, A. J., M. MUSTAFA & G. S. HENDERSON. 1988. The ecology of Sulawesi. Gadja Madah University Press, Yogyakarta. xxi + 777 pp.

WIKRAMANAYAKE, E. D., D. MARCELLINI & W. RIDWAN. 1994. The thermal ecology of adult and juvenile Komodo dragons, *Varanus komodoensis*. *In*: Abstracts, 2nd World Congress of Herpetology. pp: 284. Adelaide, Australia

WILLIAMS, E. 1957. *Hardella isoclina* Dubois redescribed. *Zool. Med.* 35(17): 235–240.

WILLIAMS, K. L. & V. WALLACH. 1989. Snakes of the world. Volume 1. Synopsis of snake generic names. Krieger Publishing Company, Malabar, Florida. viii + 234 pp.

WILLIAMSON, M. A. 1967. Notes on the growth rate of *Python reticulatus* (Serpentes: Boidae). *Herpetologica* 23(2): 130–132.

WISHAUPT, R. 1991. The mangrove snake, *Boiga dendrophila. Litteratura Serpentium* 11(2): 35–37.

WITZELL, W. N. 1983. Synopsis of biological data on the hawksbill turtle, *Eretmochelys imbricata* (Linnaeus, 1766). *FAO Fish. Syn.* (137): i–iv + 1–78.

WOLF, S. 1936. Revision der Untergattung *Rhacophorus. Bull. Raffles Mus.* 12: 137–217.

WOLFF, J. 1859. Slangen van Koetei. *Natuur. Tijd. Ned-Indië* 16(2): 206.

WONG, A. 1994. Population ecology of amphibians in different altitudes of Kinabalu Park. *Sabah Mus. J.* 1(2): 29–38.

WÜSTER, W. 1986. Taxonomic changes and toxinology: systematic revisions of the Asiatic cobras (*Naja naja* species complex). *Toxicon* 34(4): 399–406.

WÜSTER, W. 1996. The status of the cobras of the genus *Naja* Laurenti, 1768 (Reptilia: Serpentes: Elapidae) on the island of Sulawesi. *The Snake* 27(2): 85–90.

WÜSTER, W., S. OTSUKA, A. MALHOTRA & R. S. THORPE. 1992. Population systematics of Russell's viper: a multivariate study. *Biol. J. Linn. Soc.* 47: 97–113.

WÜSTER, W. & R. S. THORPE. 1989. Population affinities of the Asiatic cobra (*Naja naja*) species complex in south-east Asia: Reliability and random sampling. *Biol. J. Linn. Soc.* 36: 391–409.

WÜSTER, W. & R. S. THORPE. 1991. Asiatic cobras: systematics and snakebite. *Experientia* 47: 205–209.

WÜSTER, W. & R. S. THORPE. 1992. Dentitional phenomena in cobras revisited: spitting and fang structure in the Asiatic species of *Naja* (Serpentes: Elapidae). *Herpetologica* 48(4): 424–434.

YANG, D.-T. 1991. Phylogenetic systematics of the *Amolops* group of ranid frogs of southeastern Asia and the Greater Sunda Islands. *Fieldiana Zool.* n.s. 1423: 1–42.

YANG, D.-T. & W. LIU. 1994. Relationships among species groups of *Varanus* from southeast Asia with description of a new species from Vietnam. *Chinese Zool. Res.* 15: 11–15.

ZEIGLER, T. & W. BÖHME. 1996. Über das Beutespektrum von *Varanus dumerilii* (Schlegel, 1829). *Salamandra* 32(3): 203–210.

ZELLER, C. 1960. Das periodische Eierlegen des Klatterfrösches *Rhacophorus leucomystax* (Kuhl). *Rev. Suisse Zool.* 67: 303–308.

ZHAO, E.-M. & K. ADLER. 1993. Herpetology of China. Society for the Study of Amphibians and Reptiles, Contributions to Herpetology, No. 10, Oxford, Ohio. 522 pp + 48 pl. + 1 folding map.

ZOLLINGER, H., H. VON ROSENBERG & O. F. U. J. HUGUENIN. 1859. Reptiliën en visschen van Banjoewangi en Buitenzorg. *Natuur. Tijd. Ned-Indië* 16(2): 47.